Christian SEKIMONYO SHAMAVU
Puissant MIGISHA BIZIMUNGU

The problem of Lake Kivu gas

Christian SEKIMONYO SHAMAVU
Puissant MIGISHA BIZIMUNGU

The problem of Lake Kivu gas

comparative study with the limnic eruption of Lake NYOS in Cameroon in 1986

ScienciaScripts

Imprint

Cover image: www.ingimage.com

This book is a translation from the original published under ISBN 978-620-6-69609-4.

Publisher:
Sciencia Scripts
is a trademark of
Dodo Books Indian Ocean Ltd. and OmniScriptum S.R.L publishing group

120 High Road, East Finchley, London, N2 9ED, United Kingdom
Str. Armeneasca 28/1, office 1, Chisinau MD-2012, Republic of Moldova, Europe
Printed at: see last page
ISBN: 978-620-7-96467-3

Contents

EPIGRAPH

21 August 1986: on the evening of a full moon, in a remote valley in north-west Cameroon, almost two thousand men, women and children die. Hundreds of chickens, baboons, zebus and birds also died, while the huts and palm trees remained strangely intact.

Frank Westerman, 2015

JOB SUMMARY

The high concentrations of dissolved gases in Lake Kivu did not leave us indifferent to the catastrophe that occurred in Lake Nyos in 1986, a limnic eruption in a meromictic lake like Lake Kivu.

However, to carry out our study, we used scientific means, including objectives. The overall objective was to find out whether a limnic eruption is possible in Lake Kivu, and the specific objectives were formulated as follows: to identify the possible consequences of a limnic eruption in Lake Kivu; to draw up a plan for managing and preventing the risks and disasters of a limnic eruption in Lake Kivu; and to compare the endogenous, exogenous and heterogeneous factors between Lake Kivu and Lake Nyos with regard to a limnic eruption.

These objectives were accompanied by methods, techniques, tools and materials for collecting, analysing and interpreting the results, including the following methods: observational, phenomenological, empirical, historical, analytical, comparative, statistical and descriptive; techniques such as: documentary, interview, interview, survey and questionnaire, focus group, remote sensing, seismic, critical analysis, modelling and simulation, gravimetric, case analysis, disign, longitudinal, cross-sectional, correlation and finally the Gamma Ray technique and finally materials and tools such as: GPRS, GPS, P and S wave, compass, Telematix system, GIS, ArcGis, Seismograph and gravimeter.

We discussed the presentation, interpretation and discussion of the results, including the results of the survey, research and analysis. The results of our survey confirm our **first hypothesis** in Table 15 that the limnic eruption in Lake Kivu is possible by 100% of those surveyed. All our hypotheses were confirmed.

Clearly, given that Lake Kivu is a killer lake, this work contains proposals for a risk and disaster prevention and management plan. We have therefore formulated a project to prevent the risks of the limnic eruption of Lake Kivu.

Keywords: meromictic, limnic eruption, dissolved gas, risk and disaster

INTRODUCTION

0.1. State of the question

Lake Kivu is located in Central Africa in the East African Rift Valley (Klaus Tietze, 1981, p36) C[31] l, on the border between the Republic of Rwanda and the Democratic Republic of Congo (Kling et al., 2007, p5)[91] . It is linked to Lake Tanganyika by the Ruzizi River, which was forced to change direction by the volcanic eruptions of the Virunga chain of volcanoes. As a result, its connection with the other northern lakes was completely severed and its waters were transferred from the Nile basin to the Congo basin (Beadle, 1981, p.476)[8] .

Lake Kivu is different from other African lakes because of its tecto-volcanic origin, its altitude, its morphology and its permanent stratification due to its physico-chemical properties (Degens et al., 1973, p49)[14] . The unique character of Lake Kivu has earned it worldwide notoriety because of its inversion temperature layer, the large quantity of dissolved gases and the dangers of limnic eruptions (Jean Modeste M., 2022, p3) [74].

This lake is unusual in that its deep waters contain an enormous quantity of gas (ECHO, 2003, p5)[61] ; the deep waters of Lake Kivu in the East African Rift contain ~60 km^3 of dissolved methane and ~300 km^3 of dissolved carbon dioxide (Alfred Wüest et al., 2009, p3)[3] .

This gas is therefore one of the lake's main natural resources, but it is also a potential danger to the lake's inhabitants (ECHO, 2003, p5-6) [61]. TIETZE et al, (1999, p20)[90] , claim that Lake Kivu is the world's largest natural reservoir of methane gas and that its deep waters contain large quantities estimated at between 63 billion and 50 billion cubic metres. This is estimated at around 40 million tonnes of oil equivalent, lying under 250 m of water at the bottom of the lake.

According to Michel Halbwachs et al, (2002, p.28)[38] , the resource at stake is estimated at 65 billion m^3 or 50 million tonnes of oil equivalent. And M. Schmid et al, (2009, p15)[36] add that huge quantities of gas have accumulated there over hundreds of years; approximately ~300 km^3 of carbon dioxide (CO_2) and ~60 km^3 of methane (CH_4), is trapped in the deep waters of the lake (volume of gas at 0°C and 1 atm); and the report by Kling et al. (2007, p5)[91] , estimates that Lake Kivu could erupt, as happened in Cameroon at Lake Nyos in 1986 and Lake Monoun in 1984.

However, studies carried out by Philippe Lecrenier at the University of Liège have shown that these two gases are trapped in the deep layers of the lake (www.reflexions.uliege.be/cms/c 340052/en.les-secrets-du-lac-kivu, accessed 20 June 2023 at 11h20') [1 1.[02]

The risk of a limnic eruption in Lake Kivu must be taken into account. The lake must be degassed as quickly as possible or the methane extracted (François Misser, 2014, p81)[15] ; and ABAKIR, (2020, p63)[87] adds that, in its static state, the lake is stable and harmless. However, a major disturbance originating from the volcanic activity of Nyiragongo on the northern shore of the lake or from underwater volcanic cones could cause deep waters laden with dissolved gas to rise.

The ex-solution of CO_2 dissolved in water can occur in a very short space of time, following the supersaturation of CO_2 in the lake, which can lead to a natural disaster such as the limnic eruption at Lake Nyos (Cameroon, Africa) in 1986 (Zhiwei Cui et al., 2020 p.115)[56] .

In August 1986, in a remote region of north-west Cameroon, a carbon dioxide eruption wiped out all animal life in an instant up to twenty-five kilometres downstream from a lake of volcanic origin. It was then that we discovered that nature, whose destructive arsenal we thought we knew everything about, could also unleash a "gas war" against man (Jean-Christophe SABROUX, 2016, p1)[89] .

The concentration of CO_2 at the bottom of the lake continues to increase after the eruption, while the temperature distribution remains unchanged (Zhiwei Cui et al., 2020 p.116) [56].

0.2. Issues

According to Michel Halbwachs, Lake Kivu is potentially dangerous, containing a thousand times more dissolved gas than Lake Nyos, and could cause a gaseous emission if its stratified waters were destabilised in the same way as a limnic eruption (M. Schmid et al., 2004, p117)[38] .

To develop this study, we set out to answer the following questions:

1. Is a limnic eruption possible in Lake Kivu?
2. What would be the consequences of a limnic eruption in Lake Kivu?
3. How can the risks and disasters of the limnic eruption of Lake Kivu be managed and prevented?
4. What are the comparative endogenous, exogenous and heterogeneous factors between Lake Kivu and Lake Nyos with regard to the limnic eruption?

0.3. Assumptions

In the light of this research questioning, we will come up with the following provisional answers:

1. A limnic eruption is possible in Lake Kivu
2. The consequences of the limnic eruption would be: the death of animal species
3. Moving the population away from the high-risk area would be the best way of managing the risks; isolating the area affected by the gas emanations and trying to find the survivors would be a solution for managing the disaster; degassing Lake Kivu would prevent the risks and extracting the methane gas would prevent the disaster.
4. The comparative endogenous factors between the two lakes would be the high concentrations of dissolved gases (convergence) and that Lake Kivu contains a thousand times more than Lake Nyos before the disaster (divergence). The converging exogenous factor for Lake Kivu and Lake Nyos is the proximity of volcanic areas and the diverging exogenous factor is the flow of volcanic lava. Lake Nyos and Lake Kivu would converge on the Planck tectonic imbalance and diverge on the inland volcanic eruption.

0.4. Aim of the study

The main aim of this work is to find out whether a limnic eruption is possible in Lake Kivu.

The specific objectives of this study are to :

- Identify the possible consequences of the limnic eruption of Lake Kivu;
- Drawing up a management and disaster prevention plan for the limnic eruption of Lake Kivu;
- Comparing endogenous, exogenous and heterogeneous factors between Lake

Kivu and Lake Nyos with regard to limnic eruption

0.5. Choice and interest of subject

Our choice of this subject is justified by the fact that there is a risk of a limnic eruption in Lake Kivu.

It turns out that this study has several interests: personal, scientific, economic, philanthropic and administrative. Hence ;

> Personally, this study will enable us, as petroleum engineers, to make an effective contribution to identifying the consequences of a limnic eruption in Lake Kivu, to identify a management and prevention method for this disaster and, finally, to compare the endogenous, exogenous and heterogeneous factors between Lake Kivu and Lake Nyos with regard to the limnic eruption.

> Scientifically, this study constitutes a data bank for scientists wishing to develop studies and research in the same direction as the subject under study.

> Economically, the current exploitation of this gas could generate significant financial spin-offs that would be beneficial for the socio-economic development of two countries (DRC and Rwanda), as this study proposes the exploitation of methane, the degassing of the lake and the recovery of CO_2 as possible solutions; solutions that will generate millions in foreign currency.

> Philanthropically, our study is aimed first and foremost at the peace and safety of the population. Local residents will be able to live with the guarantee that there will be no risk of a limnic eruption.

> Administratively, the administrative authorities will then have a management and prevention plan for the limnic eruption of Lake Kivu.

0.6. Scope of the study

Our study was carried out on Lake Kivu in North Kivu in comparison with data from Lake Nyos in Cameroon in 1986.

0.7. Subdivision of work

Apart from the introduction and conclusion, this research is divided into four chapters:

- The first chapter deals with general considerations;
- The second chapter deals with the research methodology;
- The third chapter deals with the presentation and interpretation of the results.
- The fourth chapter will propose a project to prevent the risks of a limnic eruption in Lake Kivu.

CHAPTER I

General considerations

The aim of this chapter is to present the conceptual framework of the subject under study, and the presentation of our study environment, which is Lake Kivu, while describing the characteristics useful for the limnology of this lake and a brief overview of Lake Nyos in Cameroon, South Kivu and the North-West region.

1.1. Conceptual framework

1.1.1. Definitions of concepts

a. Issues :

It is the set of problems concerning a subject. (Universal Dictionary, 2010, p.1014) [8[1] 1. It is also the set of questions that a science or philosophy asks itself in relation to a given field (Petit Larousse en couleurs, 1988, p800) [83].

b. Gas :

An impalpable substance that tends to occupy the entire enclosure in which it is contained; an expandable and compressible fluid whose molecules, exerting only very weak forces on one another, can move freely relative to one another (Dictionnaire universel, 2010, p545)[81] .

c. Lake :

A large body of inland water, usually freshwater, often classified according to its origin (tectonic, glacial, volcanic, etc.) (Petit Larousse en couleurs, 1988, p570)[83] .

d. Study :

Intellectual activity by which one applies oneself to learning to know; intellectual effort applied to the acquisition or deepening of such or such knowledge. It is also a literary or scientific work on a subject that has been studied (Dictionnaire universel, 2010, p467) [811.

It is a work in which the results of research are presented (Dictionnaire Larousse junior, 2009, p388) [82].

e. Comparative :

Used to compare, involving or formulating comparisons (Dictionnaire universel, 2010, p265)[81] .

f. Comparative study :

It is a research approach that focuses on comparing similarities and differences between social phenomena in order to discover the factors or conditions that accompany the emergence of a specific social phenomenon or pattern of behaviour (Friedrich Ebert, 2016, p17)[16] .

Comparative study is also the reflective operation by which the similarities and differences between two facts are established. (J. Freyssinet- Dominjon, 2008, p13) [73].

g. Eruption :

An eruption is the violent outpouring of lava, gas or ash from the crater of a volcano (Dictionnaire Larousse junior, 2009, p378)[82] .

From the Latin "*eruptio*", from the supin of "*erumpere*", from *ex-* "out of", and *rumpere* "to break", eruption is a sudden exit or emission (of what was enclosed). (Le Grand Robert, 2005, electronic version) [84].

h. Limnique :

Limnique comes from the Greek *limnê* [marsh, lake]; applied to continental, marshy or lacustrine basins, their sediments, fauna, flora, etc. (S.A, p187)[85] . (www.lerobert.com)[106] .

i. Limnic eruption :

A limnic eruption is a type of volcanic eruption characterised by the sudden degassing of the deep waters of a meromictic lake, releasing volcanic gases (CO_2, CH_4) emitted continuously by a volcano and accumulated over many years in the deep layers of the lake (ARRICAU Victor, 2020, p16)[86] .

A limnic eruption is an explosive eruption that occurs in large volumes of stagnant water with a CO_2 solution (Zhiwei Cui et al., 2020, p115) [56].

j. Meromictic lake :

A meromictic lake is a type of lake composed of two layers of water (mixolimnion and monimolimnion). (Jean M., 2022, p1)[74] ; i.e. its waters rarely mix (Aurélien Augier, 2021, p5)[57] .

A meromictic lake is one in which the upper waters are oxygenated, unlike the anoxic lower waters. These waters never mix and are separated by a chemiocline at a depth of 60 metres (ARRICAU Victor, 2020, p16) [86] .

1.1.2. Literature review

1.1.2.1. Gas from Lake Kivu

Lake Kivu is of interest to many researchers because it is the largest natural freshwater algal digester producing biogas in the world. It is also a major reservoir of dissolved carbon dioxide and methane in East Africa (Hirslund and Morkel, 2020; and B "arenbold et al., 2020 quoted by Jean Modeste, 2022, p3) [74].

Lake Kivu is unique in the world: its deep waters contain an enormous quantity of dissolved gas. The lake is the largest reservoir of methane gas known to date (66 billion m^3 of which 55 billion m is estimated to be exploitable). "This energy manna, if exploited, would give the country an almost inexhaustible source of energy, enabling it to stop worrying about the energy needs associated with development projects" (F. BIKUMU, 2005, p4) [88]. These reserves are broken down by "data environnement" experts as shown in Table 1 (Michel Halbwachs in *data environnement*).

Table 1. Estimated methane capacity contained in each layer

	Depth	Volume of water (km)^3	Volume of methane (km)^3
IRZ	60m - 160m	176	7
PRZ	190m - 260m	138	13
URZ	260m -310m	49	16
LRZ	310m - 485m	74	30
Total		**437**	**66**

Source: Data environnement [96]

The historic programme to measure gas concentrations in Lake Kivu began in 1935 when Damas measured the concentration of carbon dioxide (CO_2) and hydrogen sulphide (H_2S). During his analysis, more than half of the CO_2 sampled was lost to the atmosphere (Damas, 1938, p125) [2°]. The next team, led by Schmitz and Kufferath, recorded the first concentrations of CH4 and CO_2 between 1952 and 1954. Unlike the first Damascus analysis, this team analysed only the quantity of degassed gas and neglected the quantity of dissolved gas remaining in the water (Schmitz and Kufferath, 1956, p56)[37] . This was followed by the measurement taken by Tietze. This was the first comprehensive study of dissolved gas concentrations and estimated the concentration of methane (CH_4) and CO_2 under standard atmospheric conditions of temperature and pressure at 30° km3 and 6° km3, respectively (Tietze, 1978, p72)[76] . Later, on the basis of updated measurements published by et al (2005), it was suggested that methane concentrations in Lake Kivu had increased by 15% since 1974. Following this suggestion, it was important that gas concentrations in the lake be managed.

The exploitable reserves in the main basin of Lake Kivu amount to 55 billion Nm^3 of methane, the equivalent of around 470 million tonnes of petrol. By exploiting this deposit at a rate of 500 million Nm^3 /year, or the equivalent of 4.25 million tonnes of petrol/year, the life of the deposit would be 110 years (F. BIKUMU, 2005, p4) [88].

Figure 1: Concentration of CO_2 and CH4 in Lake Kivu

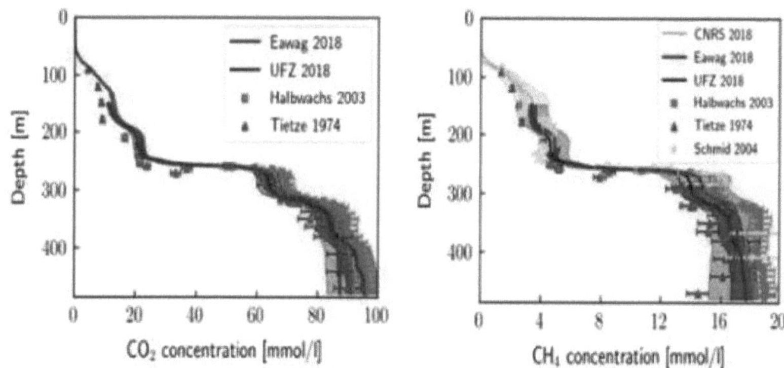

The graph on the left shows changes in vertical CO_2 profiles in Lake Kivu; the graph on the right shows changes in vertical CH4 profiles in Lake Kivu (Jean Modeste M., 2022). [74]

In view of the enormous quantities of methane and carbon dioxide, there is good reason to wonder about its origin and possible transformations.

Lake Kivu contains only methane gas and carbon dioxide, but also several other gases, as shown in Table 2:

Table 2. Different gases contained in Lake Kivu

Gas	Quantity
Carbon dioxide (CO_2) (m^3)	270 billion
Methane (CH_4) (m^3)	61 billion
Hydrogen sulphide (H_2S) (m^3)	1 billion
Nitrogen (N_2) (m^3)	10 billion
Phosphate (tonnes)	10 million
Salts (soda, potash, magnesia, lime) (tonnes)	455 million

Source: G. BORGNIEZ, 1960, p6 C I[17]

1.1.2.2. Hypotheses on the formation of CH4 and CO_2 (W. Deuser et al., 1973, p52) 541

1. Methane gas (CH_4)

Researchers (Schurtz, Kuffert, Titze et al.) proposed a hypothesis on the methane formation process. They concluded that the CH4 in Lake Kivu is produced by two processes. By reduction of magmatic CO_2 and by oxidation of organic matter by bacterial activity.

The first process contributes 2/3 of the total quantity of methane formed in the lake and is written under the chemical formula (EAWAG) :

$$CO_2+2H_2 \rightarrow CH_4+O_2 \quad (1.1)$$

One third of CH_4 is biogenic, i.e. of biological origin. This is the most likely hypothesis, assuming that this methane comes from the anaerobic decomposition of organic matter; the balance equation is given by :

$$CH_3COOH \rightarrow CO_2+CH_4 \quad (1.2)$$

2. Carbon dioxide (CO_2)

The CO_2 is produced in the lake by volcanic activity, so it comes from hydrothermal springs that are the catalysts for chemical reactions with acid rainwater infiltrating the ground. The chemical composition of these gases is exactly the same as that coming from the magma chamber and escaping to the surface through the chimney. Apart from magmatic activity, CO_2 is produced by: the decomposition of organic materials and the oxidation of methane according to the following formula:

$$CH_4+O_2 \rightarrow CO_2+2H_2 \quad (1.3)$$

Most of this methane is biogenic and recent; it would have been formed by organisms formerly classified as "methanogenic bacteria" and now reclassified as "Archaea", a group of prokaryotes distinct from truc bacteria and living in deep anoxic waters, belonging to the little-known group of *crenarchae*. These bacteria are thought to have synthesised methane from carbon dioxide and hydrogen, all of which are abiogenic (Michel Halbwachs in *data environnement and W.* Deuser et al., 1973, p52) [96] [54].

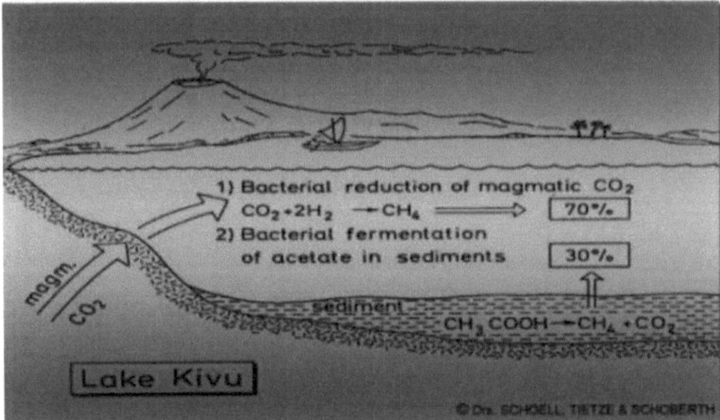

Figure 2: Origin of methane and carbon dioxide in Lake Kivu (Tietze et al.,

1981) [31].

I.1.2.3. Environmental risks

Methane and carbon dioxide are the best-known greenhouse gases (D. BENJAMIN et al., 2017, p4) [59]. The increase in their emission into the atmosphere is the cause of current global warming. CO_2 remains in the atmosphere for around a hundred years, while CH4 only stays there for around twelve years. On a century scale, however, methane is 25 times more potent than carbon dioxide in terms of global warming potential (http://inis.iaea.org, consulted on 15/06/2023) [10 7].

The precise level of risk is still the subject of analysis and discussion, but Lake Kivu is one of three lakes worldwide identified as being susceptible to serious limnic eruptions (meromictic lake); the other two are Lake Nyos and Lake Manoun in Cameroon (ECCAS and GFDRR) [58].

A limnic eruption can be caused in a number of ways: the first would be an accumulation and therefore too great a concentration of gas, which could no longer dissolve in water that was already saturated, and which would therefore rise to the surface in large quantities, in the form of bubbles. Another eruptive rise in gas could also be caused by a trigger (such as a volcanic eruption, landslide or earthquake). A fault in the exploitation process could also be a trigger (Zhiwei Cui et al., 2020, p115) [56].

This lake contains several small volcanic craters making up its bottom surface, which, once active, would create waves of internal water containing dissolved gases, the water rising to lower pressures and suddenly releasing the supersaturated gases as the pressure fell (Data environnement, consulted on 17/06/2023 at 15h00)[96] .

The more the deep water is saturated with gas, the less heat will be needed to trigger a catastrophic release of devastating gas. Global warming could be an anthropogenic source of aggravation of this risk (surface warming of 0.58°C in 30 years; but this could also be attributed to climate variability) (Lorke A., et al., 2004, p779)[6] .

There are no records of limnic eruptions in Lake Kivu. But gaps in the layers of fossil plankton at the bottom of the lake suggest that such paroxysms have struck several times over the last 5,000 years (http://svt.ac- creteil..fr/?Les-eaux-troubles-du-lac-kivu, accessed on 05/09/2023 at 3:33pm)[95] .

To promote access to electricity for most of the Congolese population and reduce demand for firewood in the east of the DRC. Environmentalists encourage the exploitation of methane gas from Lake Kivu (Kivu Power consulted on 20/08/2023 at 11:20 am) t l.[112]

I.1.2.3. Gas blocks

In 2022, the government of the Republic has launched invitations to tender for 27 oil blocks and 3 gas blocks throughout the country. All three gas blocks are located in Lake Kivu (Ministère des Hydrocarbures consulted on 16/08/2023) t l.[101]

In January 2023, the DRC government granted three companies the right to exploit the three gas blocks in Lake Kivu. These were the Americans Symbion Power & Red, for the so-called "Makelele" block, and Winds Exploration and Production LLC (Idjwi block), and the Canadians Alfajiri Energy Corporation (Lwandjofu block) (Ministry of Hydrocarbons, consulted on 20/08/2023 at 16:00)[101] .

These three gas blocks are not the only blocks in Lake Kivu. There is another block in the Gulf of Kabuno in North Kivu, which the government has submitted to Kivu Power for exploitation and degasification.

Table 3. Distribution of gas blocks on Lake Kivu

Gas block	Administrative entity	Tenderer
Makelele block	Kalehe/South Kivu territory	Symbion Power & Red
Idjwi block	Idjwi territory/South Kivu	Winds Exploration and Production LLC
Lwandjofu block	Kalehe/South Kivu territory	Alfajiri Energy Corporation
Kabuno block	Masisi/North Kivu territory	Kivu Power

Source: Sekimonyo Shamavu Christian, 2022, p23 [79]

1.2. Study environment

1.2.1. Hydrogeological environment

1.2.1.1. Lake Kivu in the DRC (Great Lakes Region)

1.2.1.1.1. Introduction to Lake Kivu

Lake Kivu, located between the DRC and Rwanda, covers an area of around 2,400km^2 . It has a maximum depth of 485m and a total volume of water of around 580km^3 . Its outlet is the Ruzizi River, which flows north of Lake Tanganyika between Uvira in DR Congo and Bujumbura in Burundi (E. DEVROEY and R. VANDERLINDEN, 1939, p3). The position of Lake Kivu is shown in Figure 3.

Figure 3: Map of Lake Kivu (Google Maps) consulted on 03/08/2023 at 21:00.

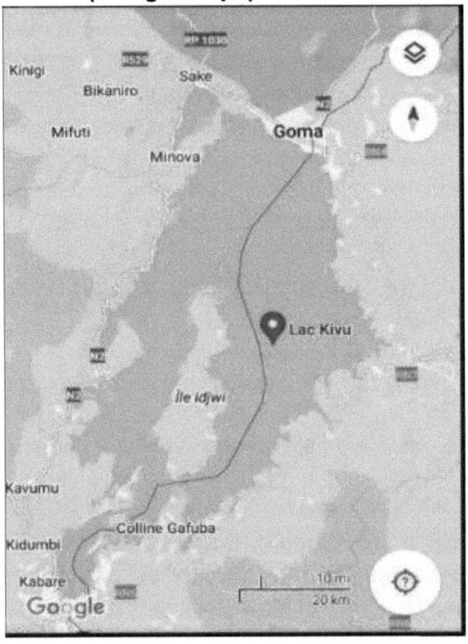

It consists of a main basin and four auxiliary basins near Bukavu, Ishungu, Kalehe and the Gulf of Kabuno, separated from the main basin by lacustrine sills. It is one of the four great lakes of the western branch of the East African Rift (Klaus Tietze, 1978, p36)[76] . It is located at an altitude of 1,465m between latitudes 1°34'30 and 2°30' South and between longitudes 28°50' and 29°25' East (Klaus Tietze, 1981, p37) [311.

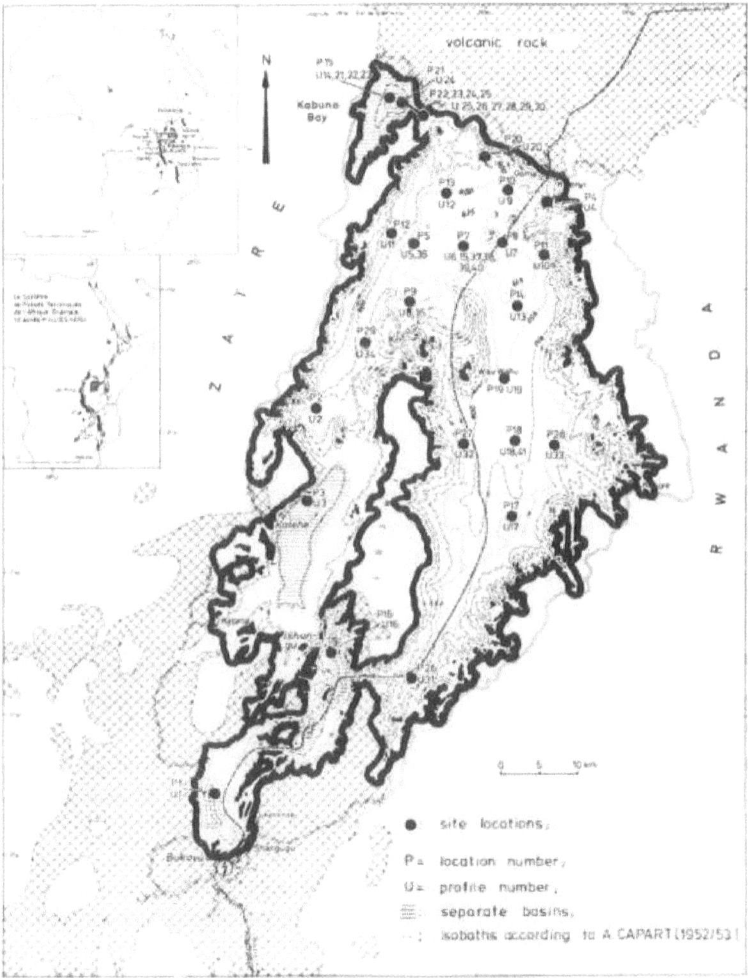

Figure 4. The Lake Kivu basins (Klaus Tietze, 1981) [27]

Lake Kivu is different from other African lakes because of its tecto-volcanic origin, its altitude, its morphology and its permanent stratification due to its chemical properties (Degens et al., 1972, p249) [1 J.[4]

Lake Kivu is one of the DRC's jewels of beauty, with its intense blue waters and lush green, finely jagged shores. Volcanoes dot the north (including the great Nyiragongo), and the salubrious climate and various livestock and crop resources attract many tourists (E. DEVROEY and R. VANDERLINDEN, 1939, p3).

With a large number of islands numbering 150, with IDJWI the largest inland island in Africa, covering an area of 315km^2 , Lake Kivu stretches from the city of Goma and the town of Sake in the north, at the bottom of the Gulf of Kabuno, to the city of

Bukavu, at the bottom of Bukavu Bay in the south; the distance as the crow flies is 106km. From east to west, the greatest width, across Mushao, is 45km. The surface area of the lake's catchment basin is 7,300km^2 including 2,600km^2 for the lake and the islands, 1,700km^2 on the eastern slopes of the Kivu massif, and 3,000km on the western slopes of the Rwandan plateaux (Gaju Gakinahe, 1991, p310)[60] .

1.2.1.1.2. Origin of Lake Kivu

Several processes can lead to the formation of lakes and ponds. Volcanic activity has given rise to certain crater lakes, and glaciation is one of the main processes responsible for the formation of cirque lakes, thaw lakes in permafrost and mill or kettle lakes. In arid regions, some lakes are the result of wind action. Rivers can be the source of oxbow lakes [...] Landslides and mudslides have led to the formation of alpine lakes. Some lakes are merely the remains of larger lakes formed during wetter prehistoric periods [...]. Beavers and humans are responsible for the formation of artificial lakes (J. Patrick DUGAN, 1997, p1213) [...].

Lake Kivu originated as a proto-lake in the mid-Pleistocene. This proto-lake is thought to be connected to the basin of ancient Lake Edward. Towards the end of the Pleistocene (25,000-20000 BC), the basin of the ancient lake should have been blocked by the accumulation of lava from the Virunga eruptions. Around 14,000 BC, this isolated basin of the lake, which had a low water level, was gradually filled, giving rise to present-day Lake Kivu. The appearance of the lake at this precise moment in history was dated by lake specialists, known as limnologists, and geological studies were undertaken (Haberyan and Hecky, 1987, p172) [22]. As a result, its connection with the other northern lakes was completely severed and its waters were cut off from the Nile basin for the Congo basin (Beadle, 1981, p476)[8] . It is one of Africa's three meromictic lakes.

1.2.1.1.3. Geological aspects

Systematic soundings, carried out between April 1935 and February 1936 by Mr Damas, thanks to subsidies granted by the Institut des Parcs Nationaux du Congo Belge and the Fonds National de la Recherche Scientifique, revealed the greatest depth of the lake, i.e. 485m. The soundings from 266 measurements showed that the bottom of the lake clearly shows the relief of an ancient valley, the slope of which gradually decreases from south to north, at the same time as the depths increase (http://reflexion.uliege.be)[102] .

The geological composition of the northern shores of Lake Kivu contains around hundreds of metres of alternating layers of solidified ash and lava produced by two active volcanoes (Nyiragongo and Nyamulagira) located near the lake (Jean Modeste, 2022, p3) [74].

The Kivu basin is bordered in the north by the chain of volcanic mountains. This volcanic dam comprises dozens of craters of varying sizes. The most majestic are, from left to right, when viewed from the lake: *Tshaninagongo* or "place of torment", currently known as Nyiragongo; it erupted between 06 December 1912 and 04 January 1913, again on 10 January 1977, again on 17 January 2002 and again on 22 May 2021, and has remained active ever since: 3,470m; *Nyamulagira*, erupts frequently: 3,056m; *Mikeno*, "the bare one": 4,437m; *Karisimbi*, named after a flanking shell used as an ornament and whose colour is reminiscent of the snow cap that tops it; it is the highest point in the range and its crater is slightly below the summit: 4,507m; *Visoke* or *Mago*, quite classic in shape: 3,711m; *Sabinyo* or *Sabyinyo* "the father with the big teeth": 3,634m; *Gahinga* "small summit": 3,474m; with two superimposed craters, and *Mihavura*, which means "landmark", with a single crater, transformed into a lake. The river of lava spewed out by Nyamulagira reached the lake on 15 December 1938, after a journey of around thirty kilometres. It was discharged into the lake at various points in the Gulf of Kabuno (André MEYER, 1955, p19-22) И.

Figure 5: Virunga volcanic chain

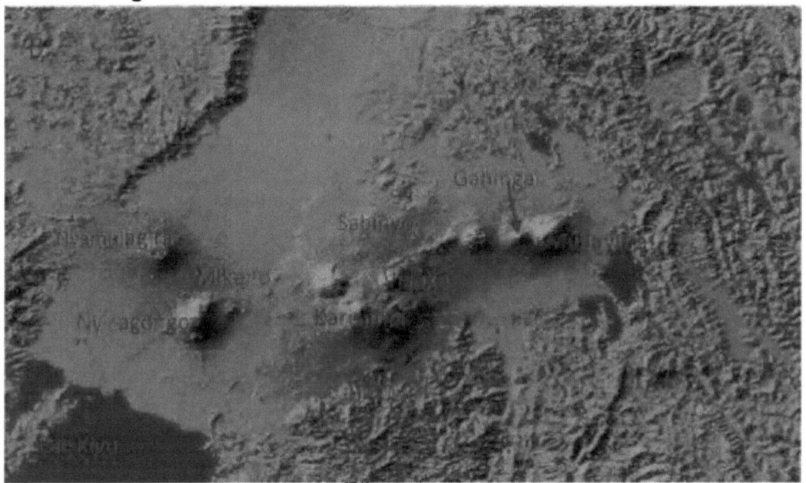

Source: Ephrem Kamate, 2018, p3 [71]

I.2.1.1.4. Bathymetry and limnimetry

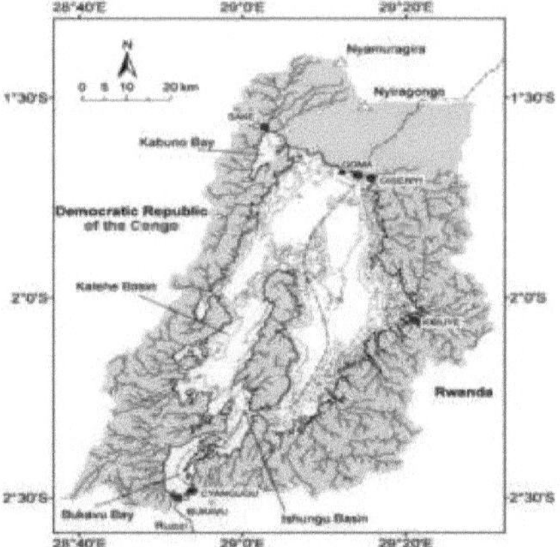

Figure 6. Bathymetric maps of Lake Kivu (The map shows the lake basins) (Alberto V. et al., 2014) [2]

Bathymetry involves measuring the depths and relief of the ocean to determine the topography of the sea bed, while limnimetry involves studying periodic variations in the height of lake levels. Recent studies of Lake Kivu by experts at *Data Environnement* have resulted in 3-D mapping of the lakebed (Data Environnement) [96].

Figure 7. Relief of the bottom of Lake Kivu (Source: data environnement) [96]

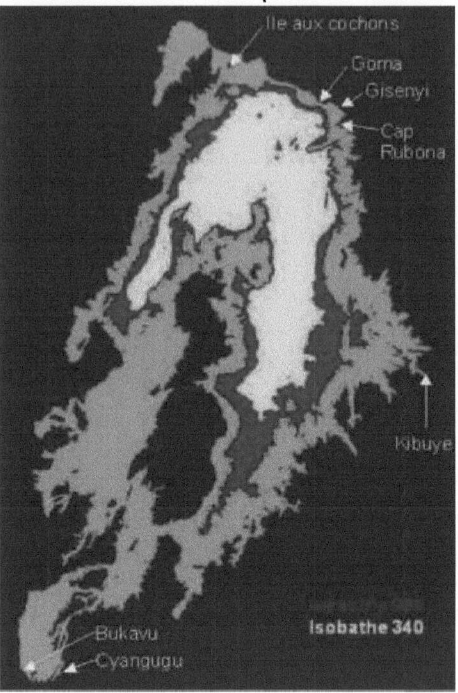

I.2.1.1.5. The climate of Lake Kivu

The climate is characterised by the distribution of dry and rainy seasons, in conjunction with the apparent movement of the sun. Average rainfall in the Kivu basin should be around 1,300mm per year. As for the regime of this lake, the water level of the lake undergoes three kinds of variations: daily variations, seasonal variations and annual variations. It appears that the variations in level are small: the total amplitude for the three-year period from January 1936 to 31 December 1938 was only 38 cm.

Solar radiation in the Kivu region has two equinoctial maxima characteristic of equatorial latitudes, plus a third corresponding to the southern solstice. This solstice coincides with perihelion (the shortest distance from the Earth to the Sun), resulting in more intense radiation. As for temperature, specific conditions of relative humidity and cloud cover mask to some extent the phenomena that are the natural consequence of the sun's seasonal movements.

There is little variation in the average daily temperature over the year: just 1° to 2°C. Depending on the station, the annual temperature varies between 16°C (Tshibinda, altitude 2115m) and 21.5°C (Katana, altitude 1500m).

As far as atmospheric currents are concerned, the prevailing winds blow from South to East and South to South-East. They rise almost every day just before noon, stirring up the surface of the water. As a result of the instability of the atmosphere above the lake, particularly at the time of the equinoxes, waterspouts can occur. They take the classic form of a double cone, rotating and translating at the same time. The permanent air currents are the northern trade wind to the north-east and the southern trade wind to the south-east.

I.2.1.1.6. Limnological characteristics

In Lake Kivu, only a limited surface layer of water is mixed. It is therefore a meromictic lake. It has a high dissolved salt content, which is reflected in high conductivity, vertical thermal stratification of the water and the presence of large quantities of methane gas, estimated at 60 billion m3 in 1978 (Tietze, 1978) [76]. The deep waters are devoid of oxygen and are topped by an oxygenated "biozone".

Surface water temperature varies very little over the year. It fluctuates between 23.1°C and 24.5°C (average 23°C). The thermal profile is uniform in the different layers of water. The temperature decreases from the surface to about 50m and then rises again in the hypolimnion to reach 25°C at 400m (Kaningini, 1995) [75] and sometimes 26°C at the deepest point of the lake.

Table 4. Limnological characteristics of Lake Kivu

Parameter	Value
Altitude (m)[(1)]	1463
Length (km) [(2)]	100
Maximum width (km)[(1)]	45
Maximum depth (m) [(1)]	485
Average depth (m) [(1)]	245
Surface area (km^2) (excluding islands) [(2)]	2370
Volume (km)[3(1)]	580
Surface area of basin (km^2) (minus lake)[(1)]	5100
Precipitation (km^3 .year)[-1(1)]	3,3
Tributaries (km^3 .year)[-1(2)]	2,0
Evaporation (km^3 .year)[-1(2)]	3,6
Outlet (km^3 .year)[-1(2)]	3,0
Temperature (°C) epilim. [(3)]	23,0-24,5
pH (3)	9,1-9,5
Transparency (m) [(3)]	3,5-6,0
Oxygen limit (m) [(3)]	70
Conductivity (µs.cm)[-1(3)]	1240
Salinity (g.L)[-1(3)]	1,115

(I)MUVUNDJA (2009) [64], (2) CHMID et al, (2010) [32] and (3) KANINGINI (1995) [ ][75]

The distribution of physical (light, heat, density, turbulence) and chemical (solute concentration) properties gives lakes a physical structure that is closely linked to their morphology, and on which the organisation of biological communities depends. As Lake Kivu is meromictic, its biozone extends to a maximum depth of 70 m, beyond which aerobic life is impossible. Only 12% of its total volume is habitable by fish (Beadle, 1981[8] and Kaningini, 1995[75]). In meromictic lakes, the deep layer, the stagnant monimolimnion, is quickly deprived of oxygen and rich in reduced chemical species (Mn^{2+}, NH_4^+, Fe^{2+}, H_2S, even CH_4).

Figure 8. Temperature and density trends in Lake Kivu

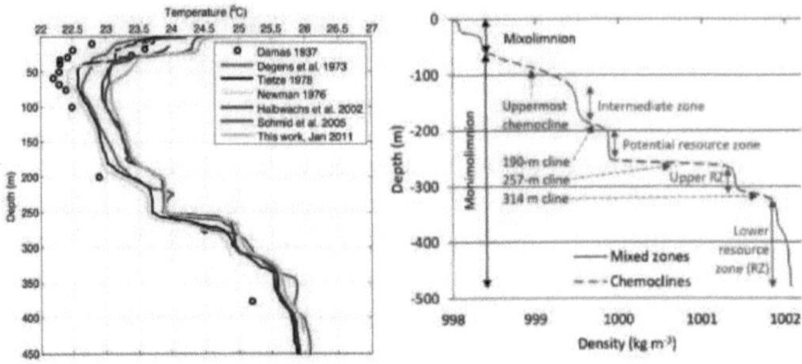

Right: profile of changes in vertical temperature and left: profile of vertical density in Lake Kivu (Jean Modeste M., 2022, p5) [74]

Lake Kivu is characterised by its physico-chemical features, in particular its high salt content of 1,115 g/L, which is reflected in high conductivity, thermal stratification of the water and the presence of large quantities of dissolved gases in the deep waters, especially CH_4, CO_2 and H_2S (Kaningini, 1995) [75].

According to Degens et al (1973)[14] , the mineral salts come mainly from hydrothermal springs emanating from the bottom of the lake and that the dissolved gas content in the water of Lake Kivu remains below saturation (salinity around 4%0). Methane, on the other hand, has a dual origin. One part is formed by the bacterial decomposition of plankton in anaerobic conditions, the other part is the result of diagenetic transformation (Tietze, 1978)[76] . The surface waters of Lake Kivu have considerable salinity and major cations are present in significant concentrations. Oxycline varies from around 30m during the rainy season to a maximum of around 65m during the dry season (Pasche et al., 2010)[44] .

Table 5. Concentrations of major ions in the surface waters of Lake Kivu (Pasche et al., 2011) [46].

Parameters	Concentration (mmol.L)$^{-1}$
Na^+	4,1
Mg^{2+}	3,8
K^+	1,9
Ca^{2+}	0,18
Alkalinity	13,3
Cl^-	0,73

All the lake basins, except Bukavu, contain a quantity of dissolved gases associated with anaerobic conditions between their various depths and the surface (Isumbisho, 2006) [25].

Damas (1935, 1937) [20] and Capart (1960) divide Lake Kivu into 5 major basins: the North basin, the Kabuno-Kashanga basin, the Ishungu basin, the Kalehe basin and the Bukavu basin (Kaningini, 1995) 5[7] .

The Bukavu basin forms the southernmost part of Lake Kivu. It is bordered to the

north-west by the Birava isthmus and to the north-east by the Nkombo and Ibindja islands. It covers an area of 96 hectares. The maximum depth of this basin is 105 m, with an average of 75 m (Kaningini, 1995) 5[7] .
The Ishungu basin is located in the southern part of the lake (2°33.94'S and 28°97.65'E) and to the north of the Bukavu basin (Isumbisho, 2000) [62]. It is around 180 m deep (Pasche, 2011)[46] .

1.2.1.1.7. Chemical characteristics of the lake

The mass of water below the coast - 275m, or 130km², contains approximately the following volumes of gas:

- 270 billion m^3 of carbon dioxide (CO_2)
- 61 billion m^3 of methane (CH_4)
- 1 billion m^3 of hydrogen sulphide (H_2S)
- 10 billion m^3 of nitrogen (N_2)
- 10 million tonnes of phosphate (PO_3)
- 455 million tonnes of various salts: soda, potash, magnesia and lime.

These volumes are calculated at conditions defined by the surface of the lake; 25° and 640mmHg (G. BORGNIEZ, 1960, p3)[17] . These gases cause a particular stratification of the lake.

1.2.1.1.8. Chemical composition of the deep waters of Lake Kivu

Nutrient enrichment of water, which originates in catchment inputs and can be relayed by mineralisation at the water-sediment interface, generally leads to an increase in the overall primary production of the ecosystem. Domestic waste and agricultural effluents are the two main sources of nutrient salts (nitrogen and phosphorus) in the catchment area.
Below 65 m, the concentration profile of most parameters in Lake Kivu was characterised by an increase with depth. The alkalinity profile is the same as for major ions, reaching a maximum level of 72.6 $mmol.L^{-1}$ at maximum depth (Pasche et al., 2010)[44] .
According to Pasche et al (2010) [44], dissolved inorganic phosphorus (DIP) and ammonium ion (NH_4^+) concentrations were high at depth (0.19 mmol/L and 4.26 mmol/L respectively). Below the oxycline, SO_4^{2-} decreases with depth to below the detection limit (<0.05mmol/L) already at 90 m (Pasche et al., 2010)[44] . The most abundant cations in Lake Kivu are Na^+ and Mg^{2+} followed by K^+ and Ca^{2+} .
N and P are essential nutrients for the growth of phytoplankton. Dead organic matter is partly mineralised and recycled at the surface. But it is during its transport in deep waters and at the water-sediment interface that organic matter is largely mineralised and nutrients are released into the water. External inputs are conditioned by atmospheric deposition, rivers and internal sources (Guyard, 2007). Around 1.0 kg of P and 0.8 kg of N per person per year are produced and deposited by human activities in the tributaries of Bukavu, ending up in Lake Kivu (Muvundja, 2010)[64] . However, current external inputs of these nutrients into Lake Kivu are still too low to cause eutrophication in less than a few decades, they add.
In the meromictic basins of Lake Kivu, N and P are mainly generated at the water-sediment interface. At this interface, 92% of N and 88% of P are mineralised and regenerated in the water column. Only 8% of N and 12% of P escape into the sediment. (Pasche et al., 2010)[44] .

1.2.1.1.9. Lake stratigraphy, density, temperature and conductivity

The variation in surface temperature over the course of the day depends on the state of the sky. On a sunny day, a variation of 27° was observed (minimum 23.4° at 7am, maximum 26.1° at 1pm, followed by a slow decrease): on a rainy day, the variation is only 1°.

From the surface to a depth of 70m, the temperature decreases. This upper zone is divided into two others: a superficial layer (*epilimnion*), 25 m thick, whose temperature varies with the seasons, and a superficial layer (*hypolimnion),* whose temperature during the year 1935 - 1936 remained fixed at 22.3°, i.e. 1° more than the average atmospheric temperature at 25m. The curves showing temperature as a function of depth therefore have a steeply sloping part (thermocline) that is more or less pronounced depending on the season: very noticeable in the rainy season (1°), it becomes less pronounced in the dry season. Every night, the temperature of the surface water falls below that of the hypolimnion, which is why the daily mixing of the upper 70m zone ensures the homogenisation of this zone.

Analysis of the gas content profiles of Lake Kivu indicates that the gas deposit is confined within the -270m isobath and that a layer favourable to the abstraction of water from Lake Kivu is located at a depth of around 350m. Analysis of the water taken from this depth shows that it contains a proportion of dissolved gas of the order of 2.5l gas/water. This gas is made up of 4/5 carbon dioxide (2.1l $_{CO2/water}$), and 1/5 methane (0.425 $_{lCH4/water}$).

A simplified description of the physico-chemical structure of Lake Kivu reveals five distinct layers:

- Biozone (BZ) ;
- *Intermediate Resource Zone (IRZ)* ;
- *Potential Resource Zone (PRZ)* ;
- Upper *Resource Zone (URZ)* ;
- Lower *Resource Zone (LRZ)* ;

Water density is a function of various physico-chemical parameters: it decreases with increasing temperature and methane content, increases with salinity (ion content characterised by the electrical conductivity parameter), and dissolved CO_2 content. The table below shows the stratification of the lake according to a number of parameters:

Table 6. Stratification of Lake Kivu

Layer	**Water (km)^3**	**Average density**	**CH4 (km)^3**	**Concentration CH4 ($_{Lmeth/Lliq}$)**	
0m - 60m	133	998,352	0	0	Biozone
60m - 190m	219	999,435	10,5	0,045	Intermediate zone
190m - 260m	84	1000,175	9,5	0,10	Potential resource
260m -310m	49	1001,447	16	0,34	Superior deposit
310m - 480m	74	1001,882	30	0,414	Lower deposit
Total	**559**		**66**		

Source: Michel Halbwachs, 17 November 2011 [63]

1.2.1.1.10. The evolution of fish and wildlife

The oxygenated zone, which occupies 70m of water depth and 12% of the total volume of the lake, is occupied by organisms. Phytoplankton and zooplankton form the fundamental basis of the fish food chain. The lake's fish fauna is relatively poor compared with other lakes in the region. It is made up of 26 species of *Clupeidae* (Isambaza) introduced to Lake Kivu in 1959 from Lake Tanganyika, 2 species of *Clariidae* (Inshonzi), 5 species of *Cyprinidae* and 18 species of *Cichlidae* (including 3 species of Tilapia and 15 species of *Haplochromis*).

Between 1958 and 1960, large quantities of larvae of the fish presumed to be *Limnothrissa miodon* and *Stolothrissa tanganyikae*, known as the Lake Tanganyika sardine, were introduced from Lake Tanganyika to Lake Kivu due to the lake's lack of pelagic species. Only *Limnothrissa miodon* has managed to develop despite its opportunistic diet. So it is clear that fishing in Lake Kivu relies mainly on *Limnothrissa miodon* and four species of *Haplochromis*, with *Cichlids*, *Tilapia* and *Clarias* playing a secondary role. There is reason to wonder about the poverty of the fauna of Lake Kivu. This can be explained by the presence of dissolved gases in the water, which hinders the development of aquatic animals at depth. That's why it would be better to focus on this gas deposit and how to exploit it.

1.2.1.1.11. Primary production and limiting nutrients in Lake Kivu

In Lake Kivu, down to about 70 m (maximum mixolimnion depth), chlorophyll a values range from 0.63 to 3µg.L⁻ 1 with an average of 1.37µg.L^{-1} .

Sarmento et al (2009)[52] measured the elemental ratios of carbon, phosphorus and nitrogen for phytoplankton in Lake Kivu and found 256.3 for C:P, 9.6 for C:N and 26.8 for N:P. Primary production in Lake Kivu was found to be strongly limited by P. This limitation would be less in Bukavu Bay, where the Si:P ratios measured are relatively low (Kaningini, 1995)[75] . Nutrient inputs to a lake are highly correlated with flood phases and are highest during the first flood. Most of the phosphorus is carried in particulate form, linked to suspended matter, and most of the nitrogen is carried in dissolved form.

Current land use in the Lake Kivu basin is dominated by subsistence agriculture using manure as fertiliser and very rarely chemical fertilisers (Muvundja et al., 2009)[42] . Deforestation in the watershed due to the need for firewood is a major problem for the future of Lake Kivu (Martineau, 2003; cited by Muvundja et al., 2009)[42] . This deforestation results in erosion and landslides (Moeyersons et al., 2004) C[40] I and the deposition of sludge in the lake.

Alongside nutrients from tributaries and atmospheric deposition, internal inputs from deep waters are the main source of phosphate to the surface waters of Lake Kivu (Muvundja et al., 2009[42] ; Pasche et al. 2010 t[44] I, Pasche et al. 2012 [[45]]). Over the centuries, the meromictic, anoxic and deep waters of Lake Kivu have accumulated both gases and nutrients (Halbwachs et al., 2002)[38] .

Table 7. Nutrient inputs to the Lake Kivu epilimnion (Muvundja et al., 2009) [64].

Nutrient intake	*SRP (t.an-1)*	*TP (t.an)^1*	*NH$^+$ (t.year-1)*	*NO3 (t.an)^1*	*TN (t.an)^1*	*SRS/ (t.an'1)*
Atmospheric deposition	118	2940	2220	1230	3450	1340
Tributaries	111	1650	370	1550	1920	23300
Internal contributions	1800	-0	18500	0	18500	29500

Phosphorus is recognised as the nutrient that controls primary production in the lake, but nitrogen can play the same role during periods of high stratification, mainly during the rainy season (Sarmentó et al. 2009)[52] .

I.2.1.1.12. Sedimentation in Lake Kivu

Seismic reflection studies reveal that unconsolidated sediments are thick in the northern basin of Lake Kivu (Wong and Herzen 1974; cited by Descy et al., 2012) [3°]. Differences in sediment thickness reflect differences in the ages of the lake basins.

In particular, the thickness is limited beyond 300 m, probably because the lake has been shallow in relation to its depth throughout history (Degens, 1973) t[1] 4l.

The materials beneath the sediments, whose thickness is unknown, are very dense; they are probably granite or metamorphic rock.

Harberyan and Hecky (1987) divided the sediment core of Lake Kivu into three different zones: zone A (14000-9400 BC): the lake was shallow with a high rate of sedimentation due to the accumulation of organic matter. The alkalinity of the lake was moderately high. The dominance of Stephanodiscus astraea indicated a low Si:P ratio. This zone was separated from zone B by a layer of ash indicating a probable internal volcanic eruption (Descy et al., 2012) [3°].

In Zone B (9400-5000 BC), the lake became deeper, with a reduced sedimentation rate. It was during this period that the Ruzizi was created as an outlet. The Si:P ratio increases during this period because of P as a limiting factor and the dominance of diatoms in the algal biomass.

Zone C, around 5000 BC, reveals dramatic changes in the lake's history. Carbonate precipitation stopped abruptly, while organic carbon and total nitrogen increased sharply. These dramatic changes have been attributed to volcanism and hydrothermal activity. (Descy et al., 2012)[30]

Haberyan and Hecky (1987)[22] believe that a limnic eruption similar to the one that occurred at Lake Nyos must have taken place at Lake Kivu, causing a mass extinction of fish. Analysis of zone C shows

that as early as 5000 BC, the lake became stratified (Haberyan and Hecky, 1987) [22]. In the early 1200s BC, Lake Kivu became meromictic because of the hot humid climate; the present stratification observed in Lake Kivu dates from this period.

The physical characteristics and sedimentation rate revealed major changes in sedimentation dating back around 50 years (Pasche et al. 2010 [44], Pasche 2012 [4[5] J). Most of the lake's physico-chemical parameters changed markedly after 1960. Since 1960, there has been a massive increase in CaCO3 and dissolved salts.

The abrupt change in the sediment core marks the start of carbonate precipitation from the 1960s onwards. In the last 50 years, the flow of organic matter has increased by 50%. More specifically, TOC has increased by 40% and TN by 80%, but TP is almost three times higher (Descy et al., 2012) [3°]. The increase in the proportion of soil in sediments reflects strong erosion in the catchment as a result of anthropogenic activities (deforestation, agriculture and mining) (Descy et al., 2012)[30]

.

The major changes 50 years ago can be explained by one or all of 3 environmental changes in Lake Kivu, i.e. changes in the trophic chain caused by the introduction of the zooplanktonophagous sardine *Limnothrissa miodon*, the high population density in the lake's catchment area, which increases external nutrient inputs. The high primary production explains the high accumulation of TOC, TN, TP and the precipitation of carbonates (Descy et al., 2012) [30].

I.2.1.1.13. Pigments

A pigment is a coloured substance, natural or artificial, of mineral or organic origin. There are also photosynthetic pigments or assimilating pigments, which are chemical compounds that enable light energy to be converted into chemical energy in organisms that carry out photosynthesis. There are two main types of photosynthetic pigment: active pigments, capable of releasing accumulated energy in three ways (fluorescence, transmission of the excited state and energy conversion) and accessory pigments, which are incapable of energy conversion (www.chem.qmul.ac.uk) 0[10] .

Active pigments

1. Chlorophyll a: blue-green pigment

Chlorophyll a is the main form of chlorophyll present in organisms that carry out photosynthesis. It is also found in small quantities in green sulphur bacteria (www.chem.qmul.ac.uk)[1 00] .

Chlorophyll a has two spectral absorption maxima in aqueous media, at around 430-440 nm in the blue and 670 nm in the red (the exact values vary depending on the composition of the solvent). It is an essential photosynthetic pigment for photosynthesis in eukaryotes, cyanobacteria and prochlorophytes due to its role as an initial electron donor in the respiratory chain (www.chem.qmul.ac.uk)[100] .

2. Chlorophyll b: yellow-green pigment

Chlorophyll b is a yellow form of chlorophyll that absorbs mainly blue light and is more soluble in aqueous media than chlorophyll a because of its carbonyl group (www.chem.qmul.ac.uk) [100].

It is not an initial electron donor in the respiratory chain but increases the energy yield of photosynthesis by increasing the amount of light energy absorbed by plants and other photosynthetic organisms. Its absorption spectrum is shifted relative to that of chlorophyll a, so that the two chlorophylls complement each other (www.chem.qmul.ac.uk) [100].

During digestive transit in phytoplankton grazers, chlorophyll pigments are progressively degraded into pheophytins (loss of Mg^{2+} from tetrapyrole), then into pheophorbides, following the action of esterases. Since pheophytins, pheophorbides and pyrophaephytins are pheopigments

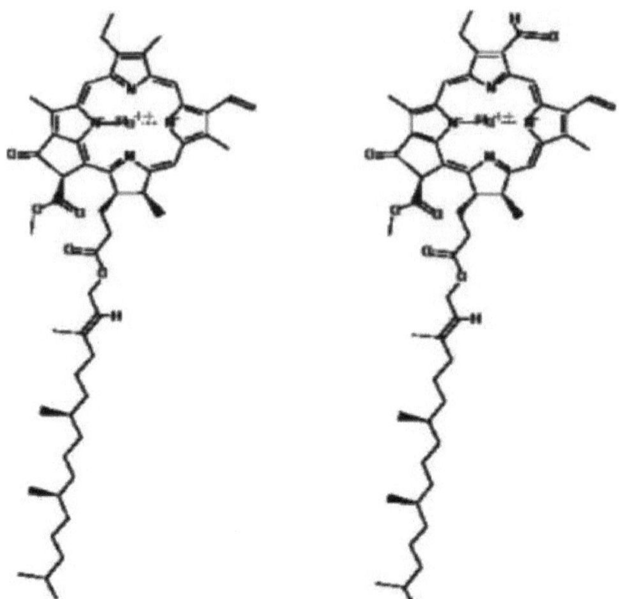

Figure 9. Structures of chlorophyll a on the left and chlorophyll b on the right [100].

3. Bacteriochlorophylls

Bacteriochlorophylls are photosynthetic pigments present in various autotrophic bacteria. They are closely related to chlorophylls, the primary pigments in plants, algae and cyanobacteria. Bacteria containing bacteriochlorophylls carry out photosynthesis, but do not produce oxygen. They use different wavelengths of light to those used for photosynthesis based on chlorophylls a and b. Bacteriochlorophylls differ depending on the bacterial group; a distinction is made between bacteriochlorophylls a to g (www.chem.qmul.ac.uk) [100].

Accessory pigments include carotene (orange pigment), xanthophyll (yellow pigment) and phycobiliproteins (water-soluble photosynthesis pigments), which in turn include allophycocyanin, phycocyanin, phycoerythrin and phycoerythrocyanin (www.chem.qmul.ac.uk) [100].

1.2.1.2. Lake Nyos in Cameroon

1.2.1.2.1. Presentation

Lake Nyos, also known as Lake Lwi, is a volcanic crater lake located in the North-West region of Cameroon, Menchum department, Wum village (Ministry of Territorial Administration, 1990, p5). It is located at an altitude of 1,091 metres, on the flank of an inactive volcano near Mount Oku, along a 1,400-kilometre-long volcanic belt known as the Cameroon Line, with Mount Cameroon (4,095 metres) remaining the only active volcano in this chain. A natural dam of volcanic rock traps the waters of the lake. The lake was once known as Lake Lwi, but was later renamed Lake Nyos, Nyos being the name of a village neighbouring the lake. Lake Nyos is located around a hundred kilometres north-west of Njindoun, the village where Lake Monoun is located (Brice Molo, 2019, p1) [10].

It lies between 06°26'23" North and 10°18'23" East. It covers an area of 1.58km^2 , is 2km long and 1.2km wide. It lies at an altitude of 1091m and is 260m deep.
Its water column can be divided into three sections separated from each other by an upper and a lower thermocline. The first layer (epilimnion) extends between 0 and -55 m where the water is mixed convectively each year during the dry season. (Paul-Alain NANA and Moïse NOLA, 2020, p81) [48] .

Figure 10. Geological map and location of Lake Nyos

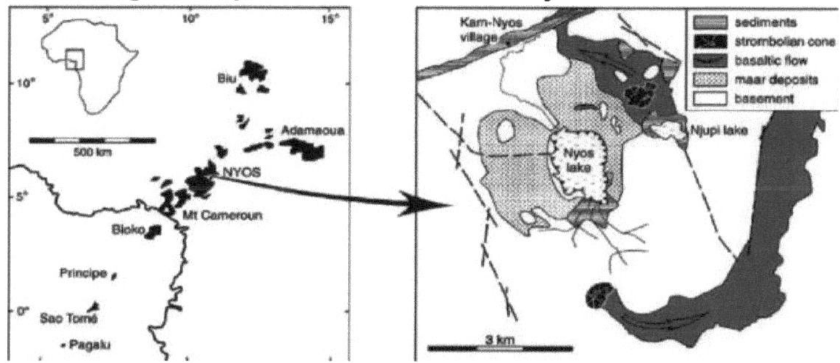

Source: R. Temdjim et al., 2004 [49]

1.2.1.2.2. The accumulation of carbon dioxide

Lake Nyos is a crater lake. As a result, carbon dioxide is [56]released at the bottom of the lake. The volume of carbon dioxide contained in the lake has been estimated at 0.3 km^3 (300 million cubic metres). Lake Nyos is located above a magma pocket. Fault lines emanate from this magma pocket and come into contact with the bottom of the lake. It is therefore an area of active volcanism.
According to some sources, Lake Nyos was formed 5 centuries ago in the crater of a dormant volcano.
Freshwater bodies can have several types of structure. Since water is a substance with a maximum density at 4°C, lakes whose surface temperature is far from this density optimum are organised into 3 thermal strata with their own physico-chemical properties, which do not mix with the other strata. On the other hand, when the surface temperature is close to 4°C, the stratification disappears and all the water in the lake can mix again by convection. In temperate zones, seasonal variations in temperature alternately stratify lakes (in summer and winter) and stir them up (in autumn and spring), oxygenating and degassing the deep waters. These are known as dimictic lakes.
In equatorial or tropical climates, the surface water temperature is always warmer than 4°C, so thermal stratification persists for years, even centuries, without any mixing taking place. These are known as meromictic lakes. In the case of Lake Lwi (Nyos), the CO_2 was therefore able to accumulate without any mixing to allow degassing. The simple fact that a gas is released into a lake is therefore not enough to create the conditions for such an accident. In addition, stratification must persist long enough for large quantities of gas to be trapped at depth.

Figure 11. Map of Cameroon's two meromictic lakes

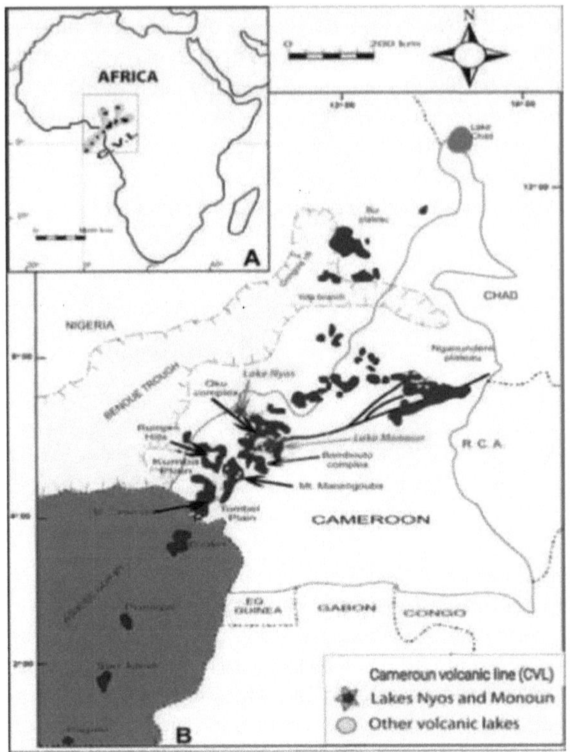

Source : Brice MOLO (2019, p4) [10]

1.2.1.2.3. Limnic eruption

A volcano, even when at rest, can continuously emit volcanic gases. When a lake is located above the point where these gases exit, as in the case of a crater lake, the gas dissolves in the water and returns to the gaseous state at the surface. If the lake is relatively deep, the volcanic gases emitted from the bottom are trapped in the lower water layers of the lake, which then accumulates these gases, sometimes for years.

According to the most common hypothesis, when a disruptive event occurs (earthquake, lake avalanche of rock debris, start of a volcanic eruption, etc.) or even when the gas concentration reaches saturation point, the water layers are inverted: bubbles of volcanic gas form in the lower layer of the lake, making it lighter and causing it to rise faster and faster towards the surface, as the system spirals out of control. The bubbles of volcanic gas then break through the surface, sometimes creating small tsunamis. When the gas is denser than air, such as carbon dioxide, which is one of the main components of volcanic gases, the gas sheet remains pressed to the ground and can flow over the crater rim along the valley floor.

In 1986, in Cameroon, a sheet of carbon dioxide emerged from Lake Nyos. Being heavier than air, the gas swept down the slopes of the volcano, killing 1,800 villagers and several thousand head of cattle in their sleep by asphyxiation (Tristan FERROIR,

2009, p9) [53].

If villages or livestock are in the path of this slick, the consequences can be dramatic: in 1986, a limnic eruption on Lake Nyos in Cameroon caused the death of more than 1,700 people and several thousand head of livestock by asphyxiation.

After a number of studies carried out by scientists on African lakes, it turns out that Lake Nyos is not the only lake concerned by a possible limnic eruption. Lake Monoun is also potentially dangerous, containing 10 million m^3 of CO_2, compared with the 300 million m^3 contained in Lake Nyos. In 1984, an eruption killed at least 37 people. A degassing operation has also been carried out on Lake Monoun since 2003.

Lake Kivu, in Central Africa, is also susceptible to such eruptions, but on a much larger scale (it has a surface area of 2,700 km^2 and several million people live on its shores).

The eruption triggered the sudden release of between 100,000 and 300,000 tonnes of carbon dioxide (CO_2). The cloud of gas initially rose at nearly 100 km/h before falling back, heavier than air, onto neighbouring villages, suffocating people and animals for 25 km around the lake.

As carbon dioxide is one and a half times heavier than air, as it escaped from the crater, it spread at ground level over a large area, reaching the surrounding villages and meadows, causing the death of villagers and their herds.

I.2.2. Administrative environment

I.2.2.1. The province of South Kivu in the DRC

I.2.2.1.1. Presentation of the Province of South Kivu

It is essential to present the Province of South Kivu by outlining its location, its political and administrative situation, its relief, its climate, its hydrographic and rainfall situation and its soil conditions.

1. Location

South Kivu Province is one of the 26 provinces that make up the Democratic Republic of Congo. It covers an area of 69,130 km^2 , or 2.78% of the national territory, and had a population of 5,772,000 in 2015, with an average density of 83 inhabitants per km^2 . The capital is Bukavu. The province lies between 28°01' east longitude and 3°01' south latitude.

The Province is limited:

- To the east, the Republic of Rwanda, Burundi and Tanzania;
- To the west, Tanganyika Province;
- In the North by the Province of North Kivu ;
- In the south, through the province of Maniema.

2. Administrative and political aspects

Since December 2006, the province of South Kivu has had the following political institutions:

- A Provincial Assembly made up of 36 members called "provincial deputies" elected by direct universal suffrage.
- A Provincial Government headed by a Provincial Governor and a Vice-Governor elected by the Provincial Assembly; and Provincial Ministers.

Administratively, South Kivu Province is divided into nine administrative entities, comprising eight territories (Fizi, Idjwi, Kabare, Kalehe, Mwenga, Shabunda, Uvira and Walungu) and the city of Bukavu. Bukavu is the capital of South Kivu Province.

The territories are subdivided into collectivities (sectors or chiefdoms) and the city into communes.

3. Relief

South Kivu is mountainous and occupies a large part of the Mitumba mountain range, with Mount Kahuzi (3,308m) as its highest point. This part also occupies the collapse ditch formed by the Ruzizi plain, Lake Kivu and Lake Tanganyika. In its western part, the Shabunda Territory forms the low-lying region extending the Maniema plateau, which slopes gently down towards the Congo River.

4. Climate

The province is located in the equatorial zone, but in its eastern part, the excesses of the climate are mitigated by the altitude. The average annual temperature is 19°C in Bukavu, 16°C in Kabare (at an altitude of 1,960m) and 10°C on Mont Kahuzi. The Shabunda and Uvira territories are the hottest regions in the province, with an average temperature of over 25°C. The dry season begins in May and ends in September.

5. Hydrography

It is abundant. There are two mountain lakes: Lake Kivu (1,470 m), which is the deepest in Africa and the second deepest in the world after Lake Baikal (1,741 m), and Lake Tanganyika (773 m). Lakes Kivu and Tanganyika are linked by the Ruzizi River. Lake Tanganyika is rich in fish. Lake Kivu, on the other hand, has very few fish due to the presence of carbon dioxide and methane.

The rivers of South Kivu belong to the Congo River basin. Most of these rivers have their source in the eastern mountains and flow westwards to the Lualaba river, while others flow into lakes.

6. Rainfall

The territories of Kabare, Walungu, Kalehe, Idjwi and the city of Bukavu have two seasons: the dry season, which lasts 3 months from June to September, and the rainy season, which lasts 9 months. During the dry season, temperatures are high and rain is scarce. This is when the swampy areas are cultivated.

The rainy season brings heavy rains, but lately, with the haphazard felling of trees, the destruction of the environment and overpopulation, the rains are becoming increasingly rare. In forest territories such as Fizi, Mwenga and Shabunda, at the entrance to the equatorial forest, it rains abundantly all year round. As for the territory of Uvira, apart from the high plateaux, rain is becoming just as scarce there, and the temperature is rising more and more because of the concentration of the population, leading to the destruction of the environment.

7. Floors

In Kabare, Idjwi and Walungu, the soil is clayey and increasingly poor due to erosion and overpopulation. As a result, there are many land disputes in this area and livestock farming is declining sharply due to a lack of pasture.

At Idjwi, the soil is still rich for farming, but the problem of overpopulation is making arable land increasingly scarce.

Kalehe also has rich clay soil, mainly due to its proximity to the forest. There are also a number of gold deposits.

The Shabunda, Mwenga and Fizi territories have sandy soils that are ideal for farming and contain significant mineral resources (gold, cassiterite, Coltan, etc.). The

Territory of Uvira also has sandy soil suitable for growing rice and cotton. Its high plateaux and very mild climate are ideal for livestock farming.

8. Methane gas

Given that the province of South Kivu borders on Lake Kivu, a lake known for its high concentrations of methane and carbon dioxide, there are 3 gas blocks in South Kivu: the Lwandjofu block, the Idjwi block and the Makele block. All these blocks were the subject of calls for tender issued by the Congolese government last year.

I.2.2.2. The North-West region

1.2.1.1.1. Introduction to the region

Northwest is one of the ten regions that make up Cameroon. Located in north-west Cameroon on the border with Nigeria, its capital is Bamenda.

The region is located in the west of the country, bordering three regions of Cameroon: South-West to the west, West to the south and Adamaoua to the east, while to the north a Nigerian state called Taraba borders the region.

The North West region lies between 6° 20' 00" North and 10° 30' 00" East. It covers an area of 17,812 km^2 and had a population of 1,728,953 in 2005. Its population density is 97 inhabitants/km^2 and its mineralogical code is NW. Its capital is Bamenda.

Figure 12. The North-West region on the map of Cameroon

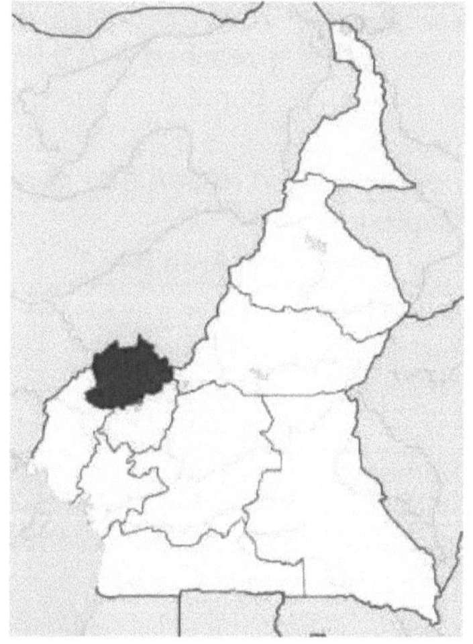

Source : Brice MOLO, 2019 [10]

1.2.1.1.2. Origin of the region

The origins of the region are linked to the settlement of the Tikar people, who joined the Bamoun Kingdom in the 1700s. In 1884, the region was colonised by Germany under the protectorate regime until 1916, when it became a condominium jointly administered by the United Kingdom and France. In 1919, administration of the North West region became solely British. In 1961, the region became part of Cameroon.

1.2.1.1.3. Administrative aspects

In 2008, the President of the Republic of Cameroon, Paul Biya, signed a decree abolishing the name "Provinces" and replacing it with "Regions".

However, in Cameroon the region is subdivided into departments and the North-West region is not spared by this administration. It is subdivided into seven departments, as shown in Figure 13.

Figure 13. Subdivision of the North-West into departments

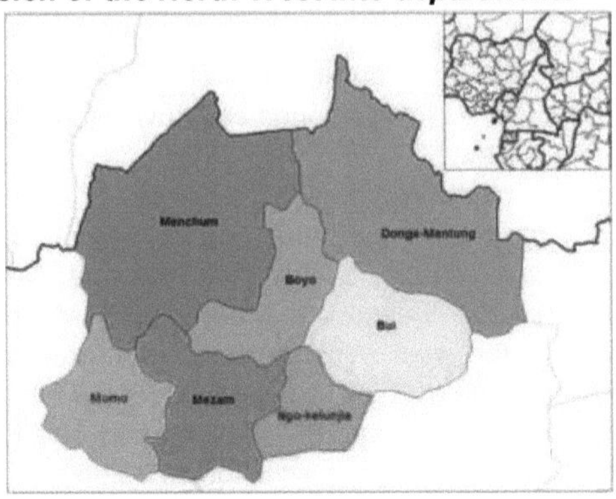

Source : R. Temdjim et al, 2004 [49]

The region, which comprises seven départements, covers an area of 17,812 km^2 and was home to over 1,840,500 people in 2001. Its population almost doubled between the 1976 and 2005 censuses, when it reached 1,728,953, slightly down on 2001 estimates. Over the same period, its population density rose from 56.7 to 99.9 inhabitants per km^2 .

Table 8. Subdivision of the North-West region into departments

Department	**Capital**	**Area (km)^2**	**Population (2001) (inhab.)**
Boyo	Fundong	1 592	169 725
Bui	Kumbo	2 297	322 877
Donga-Mantung	Nkambé	4 279	337 533
Menchum	Wum	4 469	157 173
Mezam	Bamenda	1 745	465 644
Momo	Mbengwi	1 792	213 402
Ngo-Ketunjia	Ndop	1 126	174 173

Source : Brice Molo, 2019 [J[10]

The region also has 34 arrondissements and 34 communes, as well as 559 traditional chiefdoms, including 5 first-degree chiefdoms, 117 second-degree chiefdoms and 437 third-degree chiefdoms.

1.2.1.1.4. Tourism aspect

The region contains a number of tourist sites, including chiefdoms, lakes and waterfalls. We will try to list some of these sites.

Table 9. Tourist sites in the North-West region

Chiefdoms	**Lakes**	**Falls**
Bafut chiefdom (UNESCO World Heritage site)	Lake Nyos	Menchum Falls

Mankon chiefdom ;	Lake Oku	Forest around the Menchum waterfalls
Chiefdom of Bali ;	Lake Awing	
Laikom chiefdom, near Djottin ;	Lake Awing and surrounding area	
Palace of the Fon de Nso in Kumbo (or Banso);		
Chiefdom of Mbot		

Source: Our research and analysis

1.2.1.1.5. Language

Cameroon is both a French-speaking and an English-speaking country. In the North-West region, English is spoken along with several dialects.

1.2.1.1.6. The Nyos disaster in 1986

Lake Monoun is located in the North-West region of France. It's not a very large lake, but its gas fumes managed to kill around 40 people. The media did not pay much attention to this disaster when it occurred, probably because of its "originality".

Two years later, Lake Nyos, located near Lake Monoun, experienced the same natural phenomenon, but this time with more damage than the first release from Lake Monoun.

Of recent origin, it is located on the volcanic fault line of Cameroon and was formed as a result of volcanic eruptions. The gases emitted by these lakes consisted mainly of CO_2, with very small quantities of other gases dissolved at the bottom of the lake.

Studies have shown that this phenomenon is due to an increase in water temperature and gas concentration in these lakes.

The people and livestock who died in this disaster did so by asphyxiation. The survivors had to be rehoused in overcrowded and unsanitary conditions. Cameroon was not prepared for this type of disaster, and had to call on international aid. This aid enabled the construction of resettlement areas and minimal infrastructure to meet people's needs (Tahiti BEN TCHINDA NGOUMELA, 2010) [67].

However, these measures were not enough, as the people living along the shores of Lake Nyos lived mainly from farming and livestock rearing. Most of the displaced people are still living on the resettlement sites, even though their presence there was only intended to be temporary. Genuine accompanying measures have not been taken in the long term.

Partial conclusion

In this chapter, we discussed various subjects, including the conceptual framework, where we talked about the definitions of the key concepts in relation to our subject of study; the literature review, where we talked about gas in general, from its formation to the environmental hazards it can cause; we also talked about the study environment, with the hydrogeological environment and the administrative environment. Starting with the hydrogeological environment, we outlined the quintessence of Lake Kivu and Lake Nyos, while in the administrative environment, we spoke briefly about the province of South Kivu, to which Lake Kivu seems to belong, and a brief approach to the North-West region, where Lake Nyos is located.

CHAPTER II

STUDY METHODOLOGY

SEKIMONYO SHAMAVU Christian (2023, p27) [8°] defines a research methodology as a systematic framework used to solve the research problem using the best and most feasible methods and techniques possible to conduct the research while aligning with the research goal and objectives.

In this chapter, we will present the type of research we are carrying out, the study population and, finally, the methods and techniques used to collect, analyse and process the data.

11.1. Type of search

Our work focuses on "The problem of gas from Lake Kivu: a comparative study with the limnic eruption of Lake Nyos in Cameroon in 1986".

However, our study is a comparative study. Thus, a comparative study consists of analysing and summarising the common points, differences and trends between two or more cases sharing a common interest or objective (SEKIMONYO, 2023, p62) [8°1

11.2. Study population

Our study population consisted of all the people living along the shores of Lake Kivu.

11.2.1. Sample size

Sampling is an important part of statistics. It allows us to understand what is happening in a population without having interviewed each individual, on the principle that one spoonful is enough to taste the whole soup (Javeau Claude, 1990, p43) [271. In other words, it allows us to choose a group of people to represent the others. Sampling means choosing a limited number of individuals from whom to observe and measure a characteristic in order to draw conclusions applicable to the entire population within which the choice has been made or to which it relates (DE KETELE et al., 1996, p62) t[1] 3I.

To enrich the idea, DAGNELIE (1998, p215)[11] , specifies that stratified sampling is used "...when the parent population is very heterogeneous and we wish to ensure that its different components are all represented in the sample. Stratification can then provide a significant gain in precision compared with random sampling, without changing the total number of observations to be made".

To determine the sample, we referred to the ideas of JAVEAU who said that the sample could represent 20% of the population or more for financial and temporary reasons. He also proposed taking 10% provided that the sample size is not reduced to less than 30 (1990, p45)[27] .

In view of the above, we have selected 50 petroleum engineers from the Focus des Hydrocarbures group to work throughout the Democratic Republic of Congo.

11.3. Methods used

The research method is a set of operations by which a discipline seeks to reach the truths it pursues, demonstrates them, verifies them, and above all dictates them in a more or less imperative, more or less precise, complete and systematised way (Grawitz M., 1979, p344)[18] .

In our study, we used the following methods: historical, analytical, comparative, statistical, descriptive, observational, phenomenological and empirical.

11.3.1. Data collection

Data gathering or data collection is an essential phase of an empirical study or research project during which the student gathers information that will be analysed to confirm (or not) initial hypotheses, and to answer a problem (Gaspard Claude, published on 16 December 2019 and consulted on 17/09/2023 at 22h57) [99].

Data collection can be carried out using several methods and techniques and helps the researcher to understand the phenomenon, fact or subject being studied. The methods used to collect data are observational, phenomenological and empirical.

1. **Observation method** :

It is a method of investigation which makes it possible to gather data or information about a fact, an event or about individuals (Hanhimanti, consulted on 17/09/2023 at 23h03')[103] . It is also a method of enquiry by which the researcher directly observes, through his presence in the field, the social phenomena he is trying to study (DE KETELE et al., 1990, p40)[13] . We use this method to obtain real information about Lake Kivu.

2. **Phenomenological method**

Phenomenology comes etymologically from the Greek "phainomenon" meaning "that which appears" and "logos" meaning "study". To put it plainly, it means the study of phenomena (Gildas S. NGOMO, 2016, p43)[72] . It is a method that focuses on phenomena. This method enabled us to investigate the phenomena that led to the disaster of 21 August 1986 and comparable disasters in Lake Kivu.

3. **Empirical method**

The empirical method is an approach or an intellectual conception in the sense of an abstract way or a theoretical means of discovering, in the field of investigation, envisaged realities, still hidden, on the basis of experience, i.e. observation and verification and not on the basis of theories (ANYENYOLA WELO, 2008, p220)[7] . It has helped us to verify certain truths about the environment under study.

11.3.2. In data analysis

To analyse means to break down a phenomenon in such a way as to distinguish its constituent elements. This division of an overall phenomenon into smaller elements is carried out with the aim of recognising or explaining the relationships that link these elements together, in order to better understand the phenomenon as a whole. (Mogneau P, 2008, Presses de l'Université du Québec).

This stage involved translating the raw data collected from respondents into statistical data that could be quantified and interpreted.

1. **Historical method :**

The historical method endeavours to reconstruct events right back to the source or initial fact. The historical method gathers, orders and ranks a number of facts around a single fact in order to identify the one that had the greatest influence on the fact under study (MULUMBATI NGASHA A., 2006, p18)[41] .

Using this method, we will be able to learn about the progress of studies and the possibilities of a limnic eruption in Kivu, based on the events that took place in Nyos, Cameroon, in 1986.

2. **Analytical method :**

The analytical method consists of breaking down the object of study from the most complex to the simplest. This method looks for the smallest possible component, the basic unit of phenomena (Aktouf O., 1992, p23)[1] .

This method is of great importance to us, especially as it will enable us to make significant improvements to the data collected on our subject of study.

3. Comparative method :

It is defined as a cognitive approach in which an attempt is made to understand a phenomenon by comparing different situations (Reuchlin M., 1973, p25)[50] . This method involves the researcher comparing cases that agree or differ on a single point (Stéphane Paquin, 2011, p62)[66] .

It is for this reason that the analysis of the data will identify points of convergence and divergence between these two meromictic lakes.

11.3.3. In the presentation and interpretation of results

The data analysis phase involves organising the data and breaking it down into units that can be processed and synthesised to look for patterns and trends to discover what is important and what can be learned from the data (Friedrich Ebert, 2016, p31)[16] .

Interpreting results means making sense of the results and verifying whether the hypothesis is true or false. Interpreting results involves examining the data collected through predefined processes in order to attribute meaning to the primary data and then draw a relevant conclusion (Pietro Marzo, 2023) [65].

1. Statistical method :

It is a method which attempts to reconcile qualitative and quantitative approaches, the rational and the sensory, the constructed and the observed. The aim is to quantify the qualitative and make it accessible to rigorous mathematical processing (Aktouf O., 1992, p24) ^.

It gave us the opportunity to present and interpret the results in figures and to discuss them.

2. Descriptive method :

According to N'da Paul (2002, p19)[43] , the descriptive method consists of describing, naming or characterising a phenomenon, situation or event so that it appears familiar.

This method helped us to describe the results, the study environment and to discuss the results of the survey in order to gain a better understanding of the different realities involved.

11.4. Techniques used

According to J. CHEVALIER (1992, p168) [26], research techniques are research tools involving data collection procedures adapted to the object of investigation, the method of analysis adopted and, above all, the point of view guiding the research.

A technique is defined as all the means and procedures that enable the researcher to gather the data and information necessary for his/her research subject (Universal Dictionary, 2010)[81] . For the purposes of our work, we used some of the following techniques:

11.4.1. Data collection

1. Documentary technique

Documentary technique refers to any existing source of information to which the researcher can have access. These documents can be audio, visual, audiovisual,

written or objects (P. N'da, 2002, p35)[43] .
This technique provides us with the information we need from previous studies and various documents (literature, experiments, etc.).

2. Interview technique

The interview technique is a research technique which consists of using interviews during which the researcher questions people who provide him with information relating to the subject of his research (Grawitz M., 1986, p512) [19]. This technique enabled us to talk to oil engineers in connection with the limnic eruption, where we asked them questions by word of mouth while receiving answers from them.

3. Maintenance technique

To enrich the data collected, we used the interview technique, which is a scientific investigative procedure using the process of verbal communication to gather information in relation to the set goal (Grawitz M., 1986, p586 - 587) [19].
This technique guides us in gathering information from discussions with more stakeholders in the context of our study subject.

4. Survey technique and questionnaire

Surveys cover a range of investigative methods that are technically different, but all use the declarative mode, which consists of interviewing individuals whose responses provide the information (Daniel Caumont, 2016, p72) [12].
The questionnaire consists of asking a group of respondents representative of a target population a series of questions relating to three categories of data: facts about their social, professional and family situation; subjective judgements about facts, ideas, opinions and expectations; and their knowledge of an event or problem (HOFMANN et al., 1988, p55-56) [23].
It helped us to approach our respondents and extract information from them using a survey questionnaire to be sent to oil engineers.

5. Focus group

This technique is used to gather information by presenting a group of people (between seven and twelve) with a series of statements, proposals or opinions. Each person expresses his or her opinion, and then the moderator compares the different opinions. (Patrice Stern and Jean-Marc Schoettl, 2019, p36)[47] .
This technique has enabled us to gather more, more diverse and more relevant information.

6. Remote sensing techniques

Remote sensing involves using processes and techniques to remotely acquire information about terrestrial objects, using the properties of electromagnetic waves emitted or reflected by these objects.
It is essential for gathering information about Lake Nyos.

7. Gamma Ray technique

This is a record of the natural gamma radioactivity of formations. Gamma rays are the most penetrating natural or artificial radioactive sources, and the Gamma ray spectrometer is a powerful tool for monitoring and measuring these radiations. These measurements are carried out in laboratories, on the seabed, in boreholes, on foot, on field vehicles and in the air; for decades, therefore, this method of measuring radioelement concentrations has had a wide range of applications, and significant advances have been made in calibration procedures and computer processing (software). Today, Gamma ray spectrometry is widely used in the environment, geological mapping, mining exploration and even agriculture (IAEA, 2003, p86)[24] . This technique helped us to determine the geophysical structure of Lake Kivu.

8. Seismic technology

The seismic technique is an indirect measurement technique that involves recording echoes at the surface from the propagation of an induced seismic wave underground (www.geo-ocean.fr, published on 6 May 2022, consulted on 23/09/2023)[104] . This method has helped us to gather information from seismic waves.

11.4.2. In data analysis

1. Critical analysis techniques :

The critical analysis of a text involves breaking it down into its constituent parts. It means highlighting what is clearly stated as well as revealing what is implied (S.A, 2013, p1)[78] .

This technique allows us to dissect the information received from different sources and analyse it in detail.

2. Modelling and simulation techniques

This technique consists of making systemic a mental activity preparatory to action which spontaneously remains fleeting: forming an idea, an image, a plan, a scenario of the broad lines we are going to do (Jean-Marie Van Der Maren, 2014, p239) [29].

This technique allows us to imagine what the shoreline of Lake Kivu might have been like, as if the limnic eruption had already taken place.

3. Longitudinal technique

The longitudinal technique studies phenomena over time, taking into account the different temporalities and events that coexist and/or follow one another (it can be qualitative or quantitative), in other words a long-term study (Anaelle Milon and Saeed Paivandi, 2022, p412) [4].

It compiles data from previous studies, recent studies and a view of the future of Lake Kivu.

4. Cross-disciplinary technique

The cross-sectional technique involves collecting data for different cases. With this technique, we cannot establish a causal relationship between variables but only a relationship between variables. (Scribbr published on 22/03/2018 by Justine Debret) [99]

This technique will enable us to survey a number of people of varying ages, etc.

5. Correlation technique

The correlation technique refers to a non-experimental research method that studies the relationship between two variables using statistical analysis. This technique is used to relate points of similarity.

6. Gravimetric technique

Gravimetry is a geophysical technique that measures variations in the Earth's gravitational potential field. It is a prospecting method that can be used to determine density anomalies in the subsoil (https://ground.geophysicspr.com, consulted on 23/09/2023)[113] . Gravimetry is also the study of variations in the gravity field at the surface of the ground (Dr Bouteldja Fathe, 2017, p9) [7°].

This technique helps us to identify anomalies in mass distribution in Lake Kivu.

7. The case analysis technique

Case analysis is a technique that aims to provide qualitative information through a specific study of a particular case. Intensive analysis of a unit (person or community), focusing on development factors in relation to the environment (Yves-C. GAGNON, 2012, pXI-XII) [55] .

This technique enables us to gather data to be analysed as part of our work. It also enables us to support our arguments and confirm our hypotheses, taking the case of Nyos as an example.

11.4.3. In the presentation and interpretation of results

1. Design technique

It is a technique which consists of presenting results using tables, figures, styles and/or modern graphics by computer software (Alexandra Midael, 2009, www.fnac.com consulted on 25/08/2023 at 23h50)[109] .

11.5. Tools and materials used

A tool is an object used to perform work or produce an object, and equipment is a set of objects, instruments, machines, etc. used in a service or operation of any kind, with the exception of personnel. (Le Grand Robert, 2005, electronic version)[83] .

We used a number of tools and materials, some of which are listed below:

1. GPRS

The General Packet Radio Service is an evolution of the GSM standard, which is why it is sometimes called GSM++ (or GMS 2+). It is a second-generation telephone standard enabling the transition to the third generation (3G), which is why it is called 2.5G. (University of New South Wales, https://web.maths.unsw.edu.au, consulted on 23/08/2023 at 11:23 a.m.) C[115] I.

GPRS can be used to transport voice and access data networks (particularly the Internet) using the IP or X.25 protocols.

It also allows you to :

- Point-to-point services (PTP), i.e. the ability to connect in client-server mode to a machine on an IP network,
- Point-to-multipoint services (PTMP), i.e. the ability to send a packet to a group of recipients (multicast).
- Short message service (SMS),

It was useful for us to be able to communicate information in the field.

2. THE GPS

The Global Positioning System is a US-owned utility that provides positioning, navigation and timing (PNT) services. It comprises three segments: the space segment, the control segment and the user segment. The United States Air Force develops, maintains and operates the space segment and the control segment. (https://gps.gov/systems/gps/, consulted on 23/09/2023 at 11h27) [[11] 6]. It serves and helps us locate and position wherever we may be.

3. P AND S WAVES

The sudden rupture of a fault or fault segment causes vibrations to propagate through the rock: these are known as seismic waves. When they reach the surface, these waves cause the ground to oscillate, which, depending on its amplitude and frequency, can cause damage to buildings.

The two main elastic waves generated by an earthquake are easily observed on recordings from nearby stations.
Seismic waves are elastic vibrations that can be divided into two main families: P waves, or primary waves, and S waves, or secondary waves. They are called volume waves because they propagate throughout the entire volume of the Earth. As P waves pass through, rocks are alternately compressed and relaxed like a spring. These waves propagate through all types of medium, including liquids and gases. Sound waves, in particular, are P waves. S waves, on the other hand, are shear waves: the materials they pass through are distorted and then return to their original shape. These waves can therefore only propagate in media that offer resistance to distortion, i.e. solids.
We used the seismograph to emit sound waves (P and S) in order to obtain information or data on the ground.

4. LA BOUSSOLE

The compass is a round box in which a magnetised needle oscillates, mounted on a metal pivot and dominating a dial with the compass rose. This dial is further divided, clockwise, into 360 degrees marked by a line every 5 degrees (S.A, La boussole, p2) [93].
It was useful for orienting ourselves and the map at a given location according to the cardinal points.

5. THE TELEMATIX SYSTEM

Telematics, often known as telematics systems, is the integration of telecommunications and information processing. Telematics is derived from the French word télématique, and is often used to refer to the junction of information technology and telecommunications. With the growth of the Internet and the increase in the number of telecommunications networks capable of transporting data to offices in real time for many purposes, including fleet or truck management, telematics systems have taken off more recently (https://businessyield.com, consulted on 23/09/2023 at 12h06) [117] .
We used this tool to gather information at a distance, where we couldn't reach with remote-controlled tools.

6. THE GIS

A GIS, or Geographic Information System, is a tool for importing and displaying geo-localised and statistical data for analysis on a map.
This system is used in our work to view information about Cameroon and Lake Nyos.

7. ARC GIS

ArcGIS is a complete system for collecting, organizing, managing, analyzing, communicating and disseminating geographic information. As the world's leading platform for developing and using geographic information systems (GIS), ArcGIS is used by people around the world to put geographic knowledge to work in government, business, science, education and the media. ArcGIS enables geographic information to be published so that it can be accessed and used by anyone. The system is available everywhere using web browsers, mobile devices such as smartphones and desktop computers (https://resources.arcgis.com, consulted on 23/09/2023 at 12:19 pm) [1' 4].ArcGis is software developed by Esri.
This software is very important to us in this study, as it is used, like the GIS, to obtain

geographical information for Nyos and Kivu.

8. The seismograph

A seismograph is a device used to record earthquakes, generally using the inertia of a heavy mass that tends to stay in place when the ground, to which the frame of the device is attached, moves. For very long-period waves, devices that measure the inclination of the ground are used. For very short-period waves, geophones (on land) or hydrophones (at sea) are used, which detect pressure variations (S.A., p305) [85]. This equipment is being used to gather seismological information, mainly P and S waves.

9. Gravity meter

A gravimeter is an instrument for measuring the intensity of the gravitational field (Larousse, www.larousse.fr) [1^{0} 5]. A gravimeter is in fact an accelerometer because, according to the principle of equivalence, it measures the same thing. However, a gravimeter specialises in measuring a vertical acceleration close to the Earth's normal gravity, and with a high degree of accuracy around this value. Gravimeters can take relative or absolute measurements.

Partial conclusion

This chapter deals with the research methodology, which consists of the type of research (our study being comparative), the study population and the sample size, as well as the methods and techniques used in data collection, analysis and the presentation and interpretation of the study results. Finally, several remote sensing, geological and geophysical equipment and tools were presented in this same chapter, as well as some software.

CHAPTER III

PRESENTATION, INTERPRETATION AND DISCUSSION OF RESULTS

The presentation and interpretation of the results is the stage common to all research, and is therefore the logical culmination of the process. It provides an opportunity to reflect on the theory on which the analysis is based and on the validity of its results, questions already addressed at the start of the research.

111.1. Presentation and interpretation of results

Presenting the results is a stage that involves displaying the results of the study.

Based on diagnosis and exploration, interpretation is the third objective of scientific research. Identifying a phenomenon or pattern in society and seeking sufficient information to understand it leads the researcher to provide a precise interpretation or analysis of the phenomenon studied (Friedrich Ebert, 2016, p6). [16]

111.1.2. Identity of respondents

Tableau 10. Breakdown of respondents by gender

Sex of respondents	Workforce	%	% valid	Cumulative
Male	36	75,0	75,0	75,0
Female	12	25,0	25,0	100,0
Total	**48**	**100,0**	**100,0**	

Source: Our research and analysis 2023

Interpretation: From this table, 36 engineers surveyed out of 48, i.e. 75%, are male, while 12 others, i.e. 25%, are female.

Tableau 11. Breakdown of our respondents by age

Age of respondents	Workforce	%	% valid	Cumulative
Under 18s	00	00,0	00,0	00,0
18 to 25 years old	09	18,75	18,75	18,75
26 to 30 years old	16	33,33	33,33	52,08
31 to 35 years old	18	37,50	37,50	89,58
36 to 40 years old	03	6,25	6,25	95,83
40 and over	02	4,17	4,17	**100,0**
Total	**48**	**100,0**	**100,0**	

Source: Our research and analysis 2023

Interpretation: From the above table, we can see that 18 of our respondents (37.50%) are aged between 31 and 35, 16 respondents (33.33%) are aged between 26 and 30, 09 respondents (18.75%) are aged between 18 and 25, 3 respondents (6.25%) are aged between 36 and 40, 2 respondents (4.17%) are aged 40 and over, and no respondent is aged under 18.

Table 12. Breakdown of our respondents by marital status

Marital status of respondents	Workforce	%	% valid	Cumulative
Single	22	45,83	45,83	45,83
Married	25	52,08	52,08	97,91

Widowed	01	2,08	2,08	100,0
Divorced	00	0,0	0,0	100,0
Total	**48**	**100,0**	**100,0**	

Source: Our research and analysis 2023

Interpretation: An analysis of the above results shows that 25 out of 48 respondents (52.08%) have already married, 22 (45.83%) are single, 1 (2.08%) is a widow and none of the respondents are divorced.

Table 13. Distribution of our respondents by level of education

Marital status of respondents	Workforce	%	% valid	Cumulative
Primary	00	0,0	0,0	0,0
Secondary	03	6,25	10,0	10,0
Graduated	13	27,08	26,0	36,0
Licensed	32	66,67	64,0	100,0
Total	**48**	**100,0**	**100,0**	

Source: Our research and analysis 2023

Interpretation: On the basis of this table, 32 respondents (66.67%) had a degree, 13 respondents (27.08%) had a diploma, 3 respondents had secondary education and none of the respondents had primary education.

111.1.3. Introductory question

Table 14. Knowledge of limnic eruption

Do you know about the limnic eruption?	Workforce	%	% valid	Cumulative
Yes	48	100,0	100,0	100,0
No	00	0,0	0,0	100,0
Total	**48**	**100,0**	**100,0**	

Source: Our research and analysis 2023

Interpretation: On the basis of the above, 48 respondents were aware of the limnic eruption, while none of the respondents denied any knowledge of the eruption.

111.1.4. Questions relating to the issue

Table 15. Possibility of a limnic eruption in Lake Kivu

Is an limnique eruption possible in Lake Kivu?	Workforce	%	% valid	Cumulative
Yes	48	100,0	100,0	100,0
No	00	0,0	0,0	100,0
Total	**48**	**100,0**	**100,0**	

Source: Our research and analysis 2023

Interpretation: The table above shows that 48 respondents out of 48, i.e. 100%, see a possibility of a limnic eruption in Lake Kivu.

Table 16. Consequences of the limnic eruption

What would	Workforce	%	% valid	Cumulative

consequences of the limnic eruption in Lake Kivu?				
Death of men and livestock from asphyxiation	31	64,58	64,58	64,58
Greenhouse gas emissions	09	18,75	18,75	83,33
Population displacement	02	4,17	4,17	87,50
Ecological imbalance	02	4,17	4,17	91,67
Respiratory disease and burns	03	6,25	6,25	97,92
Loss of wealth for the country	01	2,08	2,08	100,00
Total	**48**	**100,00**	**100,00**	

Source: Our research and analysis 2023

Interpretation : In this table, we note that the main consequences such as 31 of our respondents or 64.58% say the death of men and livestock by asphyxiation, 9 respondents or 18.75% speak of the emission of greenhouse gases, 3 oil engineers surveyed or 6.25% speak of respiratory diseases and burns, 2 respondents or 4.17% mention the displacement of populations, 2 others speak of ecological imbalance while only one of the respondents or 2.08% speaks of a share of wealth.

Managing and preventing risks and disasters caused by the limnic eruption

Table 17. Risk management

How can the risks and disasters caused by the limnic eruption of Lake Kivu be managed and prevented? a. How can the risks be managed?	Workforce	%	% valid	Cumulative
Move the population away from the high-risk area	23	47,91	47,91	47,91
Studying the probability of a limnic eruption	05	10,42	10,42	58,33
Living in high-altitude areas	12	25,00	25,00	83,33
Prohibiting agricultural and pastoral activities in at-risk areas	08	16,67	16,67	100,00
Total	**48**	**100,00**	**100,00**	

Source: Our research and analysis 2023

Interpretation: Based on the results found in the table above, to manage the risks, 23 of our respondents (47.91%) said that the population should move away from the high-risk zone, 12 respondents (25%) said that they should live in high-altitude zones, 8 respondents (16.67%) said that they should prohibit agricultural and pastoral activities in high-risk zones, while 5 respondents (10.42%) suggested studying the probability of a limnic eruption.

Table 18. Disaster management

How can the risks and disasters of the limnic eruption of Lake Kivu be managed and prevented?	Workforce	%	% valid	Cumulative

b. How to manage the disaster				
Insulate the area affected by the gas fumes	09	18,75	18,75	18,75
Create a management structure for disaster relief in the surrounding conurbations	07	14,58	14,58	33,33
Searching for survivors	09	18,75	18,75	52,08
Evacuating the injured	05	10,42	10,42	62,5
Setting up a crisis unit	02	4,17	4,17	66,67
Setting up a disaster assistance committee	02	4,17	4,17	70,84
Resettlement of the population in the danger zone	03	6,25	6,25	77,09
Maintain "toxic danger" signs	07	14,58	14,58	91,67
Find a remote resettlement area	04	8,33	8,33	100,00
Total	**48**	**100,00**	**100,00**	

Source: Our research and analysis 2023

Interpretation: The results obtained in the table above show that 9 of the respondents (18.75%) spoke of isolating the area affected by the gas fumes, 9 others (18.75%) said they would try to find the survivors, 7 respondents (14.58%) spoke of creating a disaster relief management structure in the surrounding towns, 7 other respondents spoke of maintaining the "toxic danger" signs, while 5 respondents (10,42% spoke of evacuating the injured, 4 respondents (8.33%) said they would look for a remote area to resettle, 3 respondents (6.25%) said they would evacuate the population in the danger zone, while 2 respondents (4.17%) spoke of creating a crisis unit and 2 other respondents of creating a committee to assist the victims; all to manage the disaster.

Table 19. Risk prevention

How can the risks and disasters of the limnic eruption of Lake Kivu be managed and prevented? c. How can risks be prevented?	Workforce	%	% valid	Cumulative
Degassing Lake Kivu	36	75,00	75,00	75,00
Installing seismic stations in the lake	09	18,75	18,75	93,75
Informing local residents of possible risks	03	6,25	6,25	100,00
Total	**48**	**100,00**	**100,00**	

Source: Our research and analysis 2023

Interpretation: This table shows that to prevent the disaster, 36 respondents (75%) were of the opinion that Lake Kivu should be degassed, 9 respondents (18.75%) suggested installing seismic stations in the lake and 03 others (6.25%) suggested informing the local population of the possible risks.

Table 20. Disaster prevention

How can the risks and disasters of the limnic eruption of Lake Kivu be managed and prevented? d. How can we prevent a disaster?	Workforce	%	% valid	Cumulative
Setting up the lake degassing project	20	41,67	41,67	41,67
Extracting methane gas	26	54,17	54,17	95,84
Regular monitoring of the lake's limnological status	02	4,16	4,16	100,00
Total	**48**	**100,00**	**100,0**	

Source: Our research and analysis 2023

Interpretation: On the basis of this table, to prevent a limnic eruption disaster, 26 of our respondents (54.17%) proposed extracting methane gas, 20 respondents (41.67%) proposed setting up a project to degas the lake and finally 2 respondents (4.16%) proposed regularly monitoring the limnological state of Lake Kivu.

Endogenous, exogenous and heterogeneous comparative factors between Lake Kivu and Lake Nyos with regard to limnic eruption

Table 21. Endogenous factors

What are the endogenous, exogenous and heterogeneous factors? between Lake Kivu and Lake Nyos in terms of the limnic eruption? a. Desfactors endogenous	Workforce	%	% valid	Cumulative
Convergence				
Both lakes are meromictic	18	37,50	37,50	37,50
High concentration of dissolved gases	23	47,92	47,92	85,42
Fuelled by volcanic gases	06	12,50	12,50	97,92
High depth	01	2,08	2,08	**100,00**
Total	**48**	**100,00**	**100,00**	
Divergence				
The gas contained in Lake Kivu is a thousand times greater than that in Nyos before the disaster.	43	89,58	89,58	89,58
The Kivu is fed by volcanic fractures, while the Nyos is located in a crater.	05	10,42	10,42	100,00
Total	**48**	**100,00**	**100,00**	

Source: Our research and analysis 2023

Interpretation: On the basis of these results, 23 respondents (47.92%) saw the high concentration of dissolved gases as an endogenous factor of convergence between Lake Kivu and Lake Nyos with regard to the limnic eruption, 18 respondents (37.50%) mentioned the fact that these two lakes are meromictic, 6 respondents (12.50%) spoke of the feeding of these two lakes by volcanic gases and only one respondent (2.08%) spoke of the depth of these two lakes as an endogenous factor of convergence.

After the respondents had mentioned the endogenous factors of convergence, it was the turn of the factors of divergence. 43 respondents (89.58%) mentioned the fact that the gas contained in Lake Kivu was a thousand times greater than that in Lake Nyos before the disaster, while 05 respondents (10.42%) spoke of the supply to Kivu by volcanic fractures and that of Nyos, which is located in an ancient crater.

Tableau 22. Exogenous factors

What are the factors Number % Valid % Cumulative
endogenous, exogenous and comparative heterogeneities between Lake Kivu and Lake Nyos with regard to the limnic eruption?
b. Exogenous factors

Convergences				
Proximity to volcanic areas	37	77,08	77,08	77,08
Earthquakes	11	22,92	22,92	100,00
Total	**48**	**100,00**	**100,00**	
Divergences				
Landslides	11	22,92	22,92	22,92
Surrounding population	18	37,50	37,50	60,42
The flow of volcanic lava	19	39,58	39,58	**100,00**
Total	**48**	**100,00**	**100,00**	

Source: Our research and analysis 2023

Interpretation: On the basis of the above table, 37 respondents (77.08%) saw the proximity of volcanic areas as an exogenous factor for convergence between Lake Kivu and Lake Nyos, while 11 respondents (22.92%) saw earthquakes as an exogenous factor for convergence.

As for divergent endogenous factors, 19 of our respondents (39.58%) mentioned volcanic lava flows, 18 respondents (37.50%) mentioned the surrounding population as an exogenous factor in the divergence between Lake Kivu and Lake Nyos, and 11 respondents (22.92%) mentioned landslides.

Tableau 23. Heterogeneous factors

What are the factors Number % valid % cumulative
endogenous, exogenous and comparative heterogeneities between Lake Kivu and Lake Nyos with regard to the limnic eruption?
c. Heterogeneous factors

Convergences				
Planck tectonic imbalance	28	58,33	58,33	58,33
Volcanic activities	20	41,67	41,67	100,00
Total	**48**	**100,00**	**100,00**	
Divergences				

Internal volcanic eruption	39	81,25	81,25	81,25
Climate change	09	18,75	18,75	100,00
Total	**48**	**100,00**	**100,00**	

Source: Our research and analysis 2023

Interpretation: According to the results of this table, 28 of our respondents (58.33%) noted that the Planck tectonic imbalance was the heterogeneous factor in the convergence between Lake Kivu and Lake Nyos, and 20 respondents (41.67%) said that volcanic activity was responsible.

For the heterogeneous divergence factors, 39 respondents (81.25%) thought that the domestic volcanic eruption was the divergence factor, while 9 respondents (18.75%) said that climate change was the divergence factor.

III.2 Discussion of results

Discussing our results means relating them to each other and to what was already known. In a metaphorical way, we could say that discussion consists in having our results converse with all the other sections (Mogneau P, 2008, Presses de l'Université du Québec).

The discussion of the results enables the results to be analysed and interpreted in relation to the research question, to which it provides some answers.

With this in mind, we conducted our fieldwork. Of our respondents, 75% were male, 37.5% were aged between 31 and 35, 52.08% were married and 66.67% had a job. All respondents said they were familiar with the limnic rash, as shown in table 14.

111.2.1. The possibility of a limnic eruption in Lake Kivu

In general, the water in the lake mixes when the surface water is cooled by the air temperatures in winter or the rivers that carry the snow melt in spring. This melt sinks and the warmer, less dense water rises from the depths of the lake, a process known as convection.

The results of our survey confirm our **initial hypothesis** in Table 15 that a limnic eruption in Lake Kivu is possible according to 100% of the people surveyed.

Scientists from the University of Liège in Belgium, in their 2014 study of Lake Kivu entitled "The Secrets of Lake Kivu", argued that too great a concentration of this gas could indeed cause a catastrophic eruption (Les secrets du lac Kivu, www.reflexions.uliege.be/cms/c 340052/en/les-secrets-du-lac-kivu, accessed on 05/09/2023 at 15h00)[102] . Two lakes in Cameroon also contain large quantities of dissolved CO_2, although in much smaller quantities than Lake Kivu. Lake Monoun experienced a gaseous eruption in 1984, and Lake Nyos, which erupted in 1986, killed nearly 2,000 people.

However, for Klaus Tietze (quoted by François Misser, 2014, p81)[15] , in the case of Lake Kivu, its particular risk of limnic eruption must be taken into account above and beyond any economic concerns.

111.2.2. The consequences of the limnic eruption in Lake Kivu

The gases dissolved at the bottom of Lake Kivu are not as docile as we know them and in the event of a limnic eruption, the negative consequences are enormous. To date, only two limnic eruptions have been observed on three lakes likely to have them (including Kivu), including Monoun in 1984 and Nyos in 1986. In the case of

Lake Monuon, the damage was ignored by the authorities and scientists until Nyos claimed almost 2,000 victims.

Based on the results of our study, the consequences of the limnic eruption of Lake Kivu would be the death of animal species (by asphyxiation), a **hypothesis confirmed** by 64.58% of our respondents (Table 16).

111.2.2.1. Positive consequences

No positive consequences are possible from the limnic eruption.

111.2.2.2. Negative consequences

- The death of men and livestock from asphyxiation:

At the time of the disaster, these deaths were reported to be the result of an explosion in the volcanic lake, which unleashed poisonous gases and fumes that asphyxiated those who breathed them. Nevertheless, the doctors who visited the region in the wake of the tragedy noted that the survivors suffered burns to the lungs and skin; clear indications of burns and injuries were evident on the corpses. These findings were not consistent with the assertion that the victims had died only from suffocation due to breathing poisoned gas (West Africa, 1987,1772-1773, p3) [68].

Mathieu YALIRE, (head of geochemistry at the Goma Volcanological Observatory) points out that it was not methane that caused the most damage, compared with carbon dioxide (CO_2). Carbon dioxide, being a heavier gas than air, would not have evaporated vertically in very high places. It would spread horizontally over the entire sub-region, causing the death of all life by asphyxiation (Auguy Blaise M., 2004, p34)[69] .

The Lake Nyos disaster remains to this day the most tragic natural event to hit Cameroon, with numerous deaths, loss of wildlife and displacement of populations. Three low-lying villages were particularly hard hit: Nyos, with around 1,200 deaths; Subum, with around 300 deaths; and Cha, with around 200 deaths. Villages far from Lake Nyos and north of Cha, in the Mbum basin, together recorded 21 deaths (Fang village, Mashi, Mundambili and Mufo) (Mesmin Tchindjang and Njilah Isaac Konfor, 2001, p52) [39].

Other consequences cited in small proportions by our respondents included greenhouse gas emissions, population displacement, ecological imbalance, respiratory illness and burns (Table 16).

111.2.3. Disaster management and prevention in the limnic eruption of Lake Kivu

On the basis of the results found in tables 17, 18, 19 and 20, 47.91% of our respondents proposed moving the population away from the high-risk area in order to manage the risks (Table 17), 18.75% proposed isolating the area affected by the gas fumes and 18.75% proposed trying to find the survivors in order to manage the disaster (Table 18).

To prevent the risks, 93.75% of our respondents proposed degassing Lake Kivu (Table 19), while 54.17% proposed extracting methane to prevent the disaster (Table 20).

111.2.3.1. Risk and disaster management

111.2.3.1.1. Risk management

Risk can be broadly defined as a situation in which there is a possibility that people or property will suffer negative consequences. It refers to the expected number of lives lost, people injured, damage to property and disruption to economic activities due to a particular natural phenomenon (CEEAC and GFDRR, 2021, p7)[58] .

According to the same ECCAS module, the risks associated with gases contrast sharply with other volcanic risks such as *lava flows, pyroclastic flows and ash falls* (p170). CO_2 is 1.5 times denser than air, it descends to low altitudes and displaces oxygen.

Moving the population away from high-risk areas, as stated by 47.91% of our respondents (Table 17), simply means that the government must take measures to identify high-risk areas and move people likely to be affected by the eruption.

111.2.3.1.2. Disaster management

A disaster is a serious disruption in the functioning of a society/community, causing widespread human, material and environmental losses, which exceed the capacity of the affected society to cope with them using its own resources (ISDR, 2017, cited by CEEAC and GFDRR, 2021, p8) [58].

However, to manage the disaster linked to the limnic eruption of Lake Kivu, our respondents suggested isolating the area affected by the gas fumes on the one hand and, on the other, trying to find the survivors, a fact attested to by 18.75% of both sides (Table 18).

1. **Insulate the area affected by the gas fumes**

Isolating the area basically means encircling or confining the zone already affected by the gases (limnic eruption) to prevent entry and exit, and then taking measures to care for survivors and evacuate those who have died.

2. **Searching for survivors**

Once the area affected by the gas fumes has been isolated, those trapped by the asphyxiating gases, victims of burns or any other obstacle, the seriously injured and survivors must always be sought out so that they can be cared for in the appropriate facilities.

Other proposals were also raised by our respondents in comparison with the limnic eruption of Nyos in 1986, such as: Creating a structure to manage disaster relief in the surrounding towns; evacuating the injured (all those who inhaled, burn victims, etc); trying to find all the survivors in the affected area; creating a crisis unit at national level; creating a provincial and/or inter-provincial committee to assist the victims; Evacuate the population from dangerous areas; Maintain "toxic danger" signs; Prohibit agricultural and pastoral activities and even construction in affected areas; Prohibit movement between 6pm and 6am; Seek a remote area for resettlement (Table 18).

111.2.3.2. Risk and disaster prevention

111.2.3.2.1. Risk prevention

It is said that "to govern is to foresee". Once the risks have been identified, it is necessary to prevent their effects. This is why our respondents proposed that 93.75% of Lake Kivu should be degassed (Table 19).

What is degassing?

This technique involves placing a polyethylene tube vertically in the lake. A mechanical pump draws in the water at the top of the column. The liquid taken from the deep waters of the lake (rich in dissolved gas) rises up the column. Its pressure decreases and the water approaches the saturation limit. When saturation is reached, bubbles begin to form and naturally rise up the column. New bubbles appear, dragging the liquid along with them. Once this process has started, the pump is no longer needed and can be switched off. A jet of water gushes out of the column and dissipates the harmless carbon dioxide into the atmosphere.

Other respondents suggested informing the surrounding population of the possible risks in order to prevent the risks associated with this type of eruption.

111.2.3.2.2. Disaster prevention

Once our results establish that a disaster linked to a limnic eruption is possible, it is imperative to prevent it. With this in mind, 54.17% of our respondents proposed extracting methane gas from Lake Kivu to prevent this disaster (Table 20). Other respondents, to a lesser extent, proposed other solutions, including a project to degas Lake Kivu and regular monitoring of the lake's limnological state;

111.2.4. Comparative endogenous, exogenous and heterogeneous factors between Lake Kivu and Lake Nyos (with regard to the limnic eruption).

111.2.4.1. Endogenous factors

We speak of an endogenous fact, a fact that comes from within (Le Grand Robert). The two lakes converge on the high concentrations of dissolved gases, as stated by 47.92% of our respondents (Table 21), while the endogenous factors of divergence are that Lake Kivu contains a thousand times more dissolved gases than Lake Nyos before the disaster by 89.58% (Table 21).

111.2.4.2. Exogenous factors

We are talking about an exogenous factor, a factor that comes from outside. The converging exogenous factor for Lake Kivu and Lake Nyos is the proximity of volcanic areas, as stated by 77.08% of our respondents, and the diverging factor is the flow of volcanic lava, for 39.58% (Table 22).

111.2.4.3. Exogenous factors

Finally, the heterogeneous factor is that which is composed of elements of a different nature. However, the comparative heterogeneous factors between Lake Nyos and Lake Kivu converge on the Planck tectonic imbalance for 58.33% and diverge on the internal volcanic eruption for 81.25% of our respondents (Table 23).

Partial conclusion

In this part, we presented the results, their interpretation and discussion of the survey results, and compared our results with the data from Lake Nyos. Our hypotheses were confirmed, and a plan for the management and prevention of risks and disasters linked to the limnic eruption was drawn up.

CHAPTER IV

PROJECT TO PREVENT THE RISK OF A LIMNIC ERUPTION IN LAKE KIVU

1. PROJECT TITLE :

The project we are proposing is entitled "Projet de prévention des risques d'une éruption limnique au lac Kivu" or P.P.R.E.L-LK for short.

2. SCOPE OF THE PROJECT

This project will be carried out on Lake Kivu.

3. PROJECT JUSTIFICATION

Our project is justified by the fact that the lake is so saturated with CO_2 that it is likely to cause a limnic eruption.

4. DURATION OF THE PROJECT

This project will be carried out over a period of 5 years or 60 months.

5. STAGES OF THE PROJECT

Table 24. Project stages

N°	STAGES OF THE PROJECT	ACTIVITIES BY STAGE
01	Decent stage at Lake Kivu	J Contacting the local authorities J Search for equipment, accommodation or camp, and other necessary requirements J Setting up teams and equipment
02	Field study stage	J Studying the Lake Kivu gas field J Studying the impact of the risk of a limnic eruption J Gather the necessary information on Lake Kivu J Learn about the animal and plant species of Lake Kivu
03	Impact assessment stage for degasification and methane extraction activities	J Reducing greenhouse gas levels on the lake J Producing fuel gas by exploiting methane gas from Lake Kivu J Spare the population from a likely limnic eruption
04	Stage in the design of risk avoidance measures	J Design a plan for degassing Lake Kivu
	ecological imbalance and the limnic eruption	J Design a responsible methane extraction plan J Please to the ecology of the lake
05	Design stage of measures to reduce limnic eruption during operations	J Drawing up a plan for setting up organs on Lake Kivu J Use appropriate equipment
06	Operationalisation stage	J Design an appropriate plan and companies for effective degassing J Signing contracts with professional

		methane extraction companies
07	**Follow-up stage for prevention measures**	**> Drawing up a plan for monitoring and evaluating the measures** **> Drawing up a plan to monitor the effectiveness, efficiency and rationality of measures** **> Drawing up an audit plan and inspecting gas companies' compliance with the measures**
08	Writing the final report	The Final Report will be written periodically over two or three days after leaving the field.

6. PROJECT SCHEDULE

Table 25. Project schedule

N°	**STAGES OF THE PROJECT**	**DURATION**
01	Decent stage in the Kabuno block	1 Month
02	Field study stage	4 months
03	Assessment of the impact of degassing activities during and after operations	2 months
04	Design stage for measures to avoid ecological imbalance and limnic eruption	2 months
05	Design stage of measures to reduce ecological imbalance and limnic eruption during degassing operations	3 Months
06	Operationalisation stage	44 Months
07	**Follow-up stage for prevention measures**	3 Months
08	Writing the final report	1 Month
	MAXIMUM PROJECT DURATION	60 months or 5 years

7. PROJECT FUNDING

Table 26. Project funding

N°	**Headings**	**AMOUNT IN USD**
01	Decent stage in the operating theatre Kabuno	USD 2,000
02	Field study stage	USD 50,000
03	Impact assessment stage degassing activities during and after the activities	USD 10,000

04	Design stage measures to avoid ecological imbalance and the limnic eruption	USD 15,000
05	Design stage measures to reduce ecological imbalance and limnic eruption during the degassing operations	USD 15,000
06	Operationalisation stage	USD 3,000,000
07	**Measurement follow-up stage**	USD 20,000
08	Report writing stage final	USD 40,000
09	Degassing loads	USD 1,200,000
10	In charge of research	USD 100,000
11	Operating load	USD 150,000
12	Expertise and consultancy	USD 150,000
13	Other expenses and contingencies	475,200 USD (20%)
AMOUNT REQUESTED FOR		USD 5,227,200
Project implementation (in figures and any letter)		Five million two hundred and twenty-seven thousand two hundred US dollars

8. TARGET PARTNER FOR PROJECT COMPLETION

The partners we are targeting for this project are :

J The Ministry of Hydrocarbons of the Democratic Republic of Congo J National and international gas companies

9. THE PROJECT'S LOGICAL FRAMEWORK MATRIX

Table 27. Project logical framework matrix

INTERVENTION LOGIC		**NARRATIVE SUMMARY OF THE PROJECT**	**Objective indicators**	**Sources or means of verification**		**Assumptions or critical conditions**	
Intervention logic	**Justification**	This project to prevent the risk of a limnic eruption in Lake Kivu is a flagship project that will	**verifiable (IOV)**	**How can you check that the measures have been successful?**		**measures**	**assumptions**
			Effective degassing of Lake Kivu and exploitation of the Kivu gas blocks			Avoidance	We would use the appropriate teams and equipment
This project has a scientific intervention	Because it applies a scientific methodology to risk prevention			For avoidance measures	Reducing greenhouse gas levels on the lake	Discount	The non-release of gases into the biozone
This	Because it			For	Producin		

project has an ecological intervention	studies the risks of a limnic eruption having 'impacts on the ecology	focus on preventing this type of scourge. The project is located to the south of Goma on Lake Kivu, and is to be carried out over a period of 60 months (5 years) at a maximum cost of USD 5,227,200.		reduction measures	g fuel gas by exploiting methane gas from Lake Kivu	would be the measure of this.
This project has an economic intervention	Cari creates a subjacent gas exploitation project on Lake Kivu					
This project has a health intervention	Because it prevents the illnesses that can result from limnic eruption					
Global objectives		The overall aim of this narrative summary is to make it easy to	Enable financial backers to verify the level of	To assess the level of performance		To make this project a reality
		ensure that the project is understood in order to obtain the funding requested	of this project			
Specific objectives		■ Show the character of the project ■ Show total cost ■ Show the duration of the project	■ Enable lessors to be fully informed about all phases of the project ■ Check	Provide the means to : ■ Enable lessors to be fully informed about all phases of the project		■ The right teams and equipment ■ No release of gases into the biozone ■ Functional degassing installations

		the level of execution ■ Assessing the purpose of the project	■ Check the level of execution ■ Assessing the purpose of the project	
Results	The project understood and the financing obtained	The project carried out and evaluated	Satisfactory results, reduced risks	The project was well executed
Activities	■ Communicating the nature of the project ■ Show total cost ■ Respecting the project duration	■ Make all phases of the project fully known to the funding bodies ■ Monitor the project frequently ■ Assessing the purpose of the project	■ Informing financial backers about all phases of the project ■ Draw up a project monitoring report ■ Evaluating the project	■ Providing qualified teams with appropriate equipment ■ Releasing harmless gas into the atmosphere ■ Regular operation of degassing facilities

10. EXPECTED RESULTS

- **Preventing the risk of a limnic eruption**
- Operating degassing facilities
- Ensure that activities do not weaken the monimolimnion
- The ecological impact during and after operations must be taken into account

11. PROJECT SUMMARY

This project, entitled "Project to prevent the risks of a limnic eruption in Lake Kivu", has several aims: from a scientific point of view, it introduces a scientific methodology for risk prevention; from an ecological point of view, it studies the risks of a limnic eruption having an impact on the ecology; from an economic point of view, it creates a subtle gas exploitation project on Lake Kivu; and from a health point of view, it prevents illnesses that could be caused by a limnic eruption and preserves the health of those who could lose their lives as a result.

GENERAL CONCLUSION AND SUGGESTIONS

Lake Kivu, situated in the East African Rift in a spectacular volcanic landscape, has fascinated the local population and inspired legends; nineteenth-century explorers, inspiring romantic reports; and twentieth- and twenty-first-century scientists, inspiring limnological and geochemical research. For some, Lake Kivu is a "killer lake", containing large quantities of carbon dioxide and methane in its deep, anoxic waters, and it has been compared to Lakes Nyos and Monoun, whose eruptions have caused massive animal and human mortality in Cameroon (J.-P. Descy et al., 2012, p1)[30] .

Several studies have been carried out on Lake Kivu and Lake Nyos since colonial times, and this work follows on from them. The high concentrations of dissolved gases in Lake Kivu did not leave us indifferent to the incident that occurred in Lake Nyos in 1986.

This study focused on comparing the various comparable dividends between Lake Nyos before, during and after the Lake Nyos disaster with the current characteristics of Lake Kivu in terms of limnic eruption.

To achieve this, we set ourselves a number of objectives, the overall objective being to find out whether a limnic eruption is possible in Lake Kivu, and the specific objectives were formulated as follows:

- Identify the possible consequences of the limnic eruption of Lake Kivu;
- Drawing up a management and disaster prevention plan for the limnic eruption of Lake Kivu;
- Comparing endogenous, exogenous and heterogeneous factors between Lake Kivu and Lake Nyos with regard to limnic eruption

To carry out this study, we used a number of methods, techniques, tools and materials to gather, analyse and interpret the results, including observational, phenomenological, empirical, historical, analytical, comparative, statistical and descriptive methods; techniques such as : documentary, interview, interview, survey and questionnaire, focus group, remote sensing, seismic, critical analysis, modelling and simulation, gravimetric, case analysis, disign, longitudinal, cross-sectional, correlation and finally the Gamma Ray technique and finally materials and tools such as: GPRS, GPS, P and S wave, compass, Telematix system, GIS, ArcGis, Seismograph and gravimeter.

This study led to the following conclusions: The results of our survey confirm our **first hypothesis** in Table 15 that a limnic eruption in Lake Kivu is possible according to 100% of the people surveyed (1); the consequences of a limnic eruption in Lake Kivu would be the death of animal species (by asphyxiation), a **hypothesis confirmed** by 64.58% of our respondents (Table 16); according to the results found in tables 17, 18, 19 and 20, to manage the risks, 47.91% of our respondents proposed moving the population away from the high-risk area (Table 17), 18.75% proposed isolating the area affected by the gas fumes and 18.75% proposed trying to find the survivors in order to manage the disaster (Table 18); to prevent the risks, 93.75% of our respondents proposed degassing Lake Kivu (Table 19), while 54.17% proposed extracting methane gas from Lake Kivu to prevent the disaster (Table 20).

The two lakes converge on high concentrations of dissolved gases, as stated by

47.92% of our respondents (Table 21), while the endogenous divergence factor is that Lake Kivu contains a thousand times more dissolved gases than Lake Nyos before the disaster, as stated by 89.58% (Table 21). The converging exogenous factor for Lake Kivu and Lake Nyos is the proximity of volcanic areas, as stated by 77.08% of our respondents, and the diverging factor is the flow of volcanic lava (39.58%) (Table 22). Finally, the heterogeneous factor is made up of elements of a different nature. However, the comparative heterogeneous factors between Lake Nyos and Lake Kivu converge on the Planck tectonic imbalance at 58.33% and diverge on the inland volcanic eruption at 81.25% of our respondents (Table 23).

Finally, that the government can avoid a disaster similar to the one experienced in the North-West (Cameroon) on 21 August 1986 in Nyos, which cost the lives of many people and livestock, as well as an irreversible ecological disaster that would cost the lives of more than 5 million people in the surrounding areas (Goma, Bukavu, Idjwi, Kalehe, etc.). With this in mind, we have drawn up a project to prevent the risks associated with the limnic eruption of Lake Kivu for the government to use.

With this in mind, it would be important in future studies to look at the implementation of the risk and disaster prevention and management plan as found in this study. Any contribution to improving this work is essential.

BIBLIOGRAPHY

1. GENERAL WORKS

1. AKTOUF Omar, (1992: 2ème edition), *Méthodologie des sciences sociales et approche qualitative des organisations*, Canada (Québec): Presses de l'Université du Québec; 213pages. ISBN 2760504573, 9782760504578

2. Alberto V. Borges, Cédric Morana, Steven Bouillon, Pierre Servais, Jean-Pierre Descy and François Darchambeau (2014) ; *Carbon cycling of Lake Kivu (East Africa) : Net Autotrophy in the Epilimnion and Emission of CO2 to the Atmosphere Sustained by geogenic Inputs ;* Ed. PloSOne, 9(10): e1095000, Geophysical research Abstracts; 196 pages, DOI 10.1371/journal.pone.0109500

3. Alfred Wüest, Lukas Jarc and Martin Schmid, (2009); *Modelling the reinjection o deep-water after methane extraction in lake Kivu*, Switzerland (Kastanienbaum): EAWAG, Research and ManagementCH-6047 Kastanienbaum; 141pages

4. Anaelle Milon and Saeed Paivandi (2022) ; *Enquêter dans les métiers de l'humain : Traité de méthodologie de la recherche en sciences de l'éducation et de formation, Tome I*, France (Loraine) : Editions Raison et Passions : Hors collection, 286 pages

5. André MEYER, (1955), *Aperçu historique de l'exploration et de l'étude des regions volcaniques du Kivu, parc National Albert,*VIII, Missions d'études Voulcanologiques, Fascicule 1, Congo-Belge (Léopoldville).

6. Andreas Lorke, Klaus Tietze, Michel halbwachs, Alfred Wüest, (2004); *Response of Lake Kivu stratification to lava inflow and climate warming,* Switzerland (Kastanienbaum): American Society of Limnology and Oceanography, Inc; 49 (3), pp778 - 783, 6pages

7. ANYENYOLA WELO (2008 :2è édition) ; *Essai de sociologie de la religion*, RDC (Lubumbashi) : Presses Universitaires de Lubumbashi (UNILU)

8. Beadle Leonard C., (1981: 2nd edition); *The inland waters of tropical Africa - An introduction to tropical limnology*, England (London), Longman publishers, 475pages

9. Bertram Boehrer, Wolf von Tümpling, Ange Mugisha, Christophe Rogemont, Augusta Umutoni (2020) ; *Reliable reference for the methane concentration in Lake Kivu at the beginning of industrial exploitation*, Rwanda, 22p

10. Brice MOLO, (2019) ; *une géohistoire des catastrophes rumourogènes au Cameroun : les éruptions limniques de Njidoun et Nyos 1984 - 1986*, Géocarrefour

11. DAGNELIE P., (1998) ; *Statistique théorique,* Tome1, éd. De Boeck, Bruxelles

12. Daniel Caumont (2016), *Les études du marché*, Ed. Dunod, collection Les topos, 5ème edition, Paris, ISBN 978-2-10-074548-7

13. DE KETELE J. M. et ROEGIERS X., (1996) ; *Méthodologie de recueil d'information Fondement des méthodes d'observation, de questionnaires, d'interview et d'étude de documents*, De Boeck, Bruxelles

14. Degens et al, (1973); *Lake Kivu : Structure chemistry and biology of an East African Rift Lake*, Geologisch Rundschaer, n° 62, 245p

15. François Misser, (2014) ; *Hydrocarbures : l'Etat affirme sa volonté d'exploiter la ressource, Conjonctures congolaises*, 65 - 96, 32p

16. Friedrich Ebert Stiftung, (2016), *Scientific research methodology for civil society organisations*, 50p

17. G. BORGNIEZ, (1960 : fasc. 1); *Données pour la mise en valeur du gisement de méthane du lac Kivu ;* mémoires in 8°, nouvelle série, Tome XIII, 1960, Bruxelles 5, 117p
18. Grawitz Madeleine (1979:4ème ed.); *Méthodes des Sciences Sociales*, Dalloz, Paris
19. Grawitz Madeleine (1986:3ème ed.); *Méthodes des Sciences Sociales*, Dalloz, Paris
20. H. Damas (1938), *Exploration du Parc National Albert, Institut des Parcs nationaux du Congo Belge*, Mission H. Damas (1935 - 1936), Ed. LELOUP, Brussels
21. H. Damas (1938), *Quelques caractères écologiques de trois lacs équatoriaux : Kivu, Edouard, Ndalaga ;* Tome LXVIII, Institut Ed. van Bereden, Bruxelles, Imprimérie Forton, Victor Greyson, pp121-135, 16p
22. Haberyan K. A. and Hecky R.E., (1987), *The late Pleistocene and Holocene stratigraphy and Palelimnology of Lakes Kivu and Tanganyika, Paleogeography, Paleoclimatology, Paleoecology*; 61, pp169-197
23. HOFMANN A. M. et BRAY L., (1988), *Le travail de fin d'études : une approche méthodologique du mémoire*, Ed. Masson, Paris
24. IAEA, (2003); *GUIDELINES FOR RADIOELEMENT MAPPING USING GAMMA RAY SPECTROMETRY DATA*, International Atomic Energy Agency, IAEA, VIENNA, IAEA-TECDOC-1363, ISBN 92-0-108303-3, 173p.
25. Isumbisho M., Sarmento, H., Kaningini, B., Micha J.C. and Descy, J.P., (2006); *Zooplankton of Lake Kivu, East Africa: half a century after the Tanganyika sardine introduction*. J. Plankton Research 28(11): 1 -989.
26. J. CHEVALIER (1992); *Administration de l'entreprise*, Ed. Dunod, Paris
27. Javeau Claude (1990 :4ème éd.); *L'enquête par questionnaire : Manuel à l'usage du praticien*, 2e tir., Edition de l'université de Bruxelles
28. Jean Verbeke (1957) ; *Recherches écologiques sur la faune des grands lacs de l'Est du Congo Belge*, Brussels, 215p
29. Jean-Marie Van Der Maren (2014 :3ème edition), *applied research for professionals*, Ed. De Boeck, 304p
30. Jean-Pierre Descy, François Darchambeau and Martin Schmid, (2012); *Lake Kivu : Limnology and biogeochemistry of a tropical great lake*, Acquatic Ecology Series 5, Springer Science + Business Media BV, 196p
31. Klaus TIETZE, (1981); *Direct Measurements of the In-Situ Density of Lake and Sea Water with a New Underwater Probe System*, Geophysica, Vol 17, No 1-2, p.33-45.
32. Klaus Tietze, Finn Hirslund, Philip Morkel and John Boyle, Alfred Wüest and Martin Schmid (2010); *Management prescriptions for the Development of the Lake Kivu Gas Resources,* Expert working Groupon Lake Kvu Extraction, Final version for General Release, (DRC - Rwanda), 38p
33. Klaus Tietze, Geyh M., Müller H. and Schröder L. (1980); *The genesis of methane in Lake Kivu (Central Africa)*, Geol. Rundsch. 69, 452 - 472.
34. Kling G. W., Clark M. A., Compton H. R., Devinee J. D., Evans W. C., Humphrey A. M., Koeningsberg E. J., Lockwood J. P., Tuttle M. L. and Wagner G. N., (1987); *The 1986 Lake Nyos disaster in Cameroon,* West Africa, Science, 236, , 169-175, 6p
35. M. POLL and H. Damas, (1938 : Fascicule 6); Poissons, *Parc national Albert,*

mission H. Damas 1935 - 1936, 96p

36. M. Schmid, Alfred Wüest and Lukas Jarc, (2009); *Modelling the reinjection of deepwater after methane extraction in Lake Kivu*, EAWAG, Seestrasse 79, CH-6047 Kastanienbaum, Switzerland, 141p

37. M. Schmittz and Kufferath, (1956); *Recherche au Congo Belge et leur résultat pratique, bulletin agricole du Congo*, Brussels.

38. Martin Schmid, Andreas Lorke, Daniel Frank McGinnus, Michel Halbwachs, Klaus Tietze and Alfred Wüest, (2004); *How hazardous is the gas accumulation in Lake Kivu? Arguments for a risk assessment in light of the Nyiragongo volcano eruption of 2002*, pp115-122, 8p

39. Mesmin Tchindjang and Njilah Isaac Konfor, (2001) ; *Risque d'inondation dans la vallée de Nyos*, Laboratoire de Géomorphologie, African Journal of Science and Technology, Université de Yaoundé, vol. 2, n°2, pp50-62, 13p

40. Moeyersons, J., Tréfois, P., Lavreau, J., Alimasi, D., Badriyo, I., Mitima, B., Mundala, M.,Munganga, D.O., and Nahimana, L., (2004) ; *A geomorphological assessment of landslide origin at Bukavu, Democratic Republic of the Congo;* Elsevier, Engineering Geology 72: 73 - 87.

41. Mulumbati Ngasha A., (2006); *Introduction à la science politique*, Ed. Africa, Lubumbashi, 2006, 543p

42. Muvundja F., Pasche N., Bugenyi F. W.B., Isumbisho M., Namugize B., Rinta P., Schmid M., Stierli R., Wüest A., (2009). *Balancing nutrient inputs to lake Kivu, Journal of great lakes researchs*, INS 0 380-1330.8. DOI : 10.1016/j.jglr.2009.06.02

43. N'da P., (2002 :2ème édition) ; *Méthodologie de la Recherche, de la problématique à la discussion des résultats,* Edition universitaire de Côte d'Ivoire, Abidjan.

44. Pasche N, Alunga G, Mills K, Muvundja F, Ryves DB, Schurter M, Wehrli B, Schmid M., (2010.) *Abrupt onset of carbonate deposition in Lake Kivu during the 1960s: response to recent environmental changes*. J. Paleolimnol. 44: 931-946.. DOI 10.1007/s10933-010-9465-x

45. Pasche N., Muvundja A. F., Schmid M., Wüest A. and Müller B. Descy J.-P.,(2012). (eds.), *Lake Kivu: Limnology and biogeochemistry of a tropical great lake, Aquatic Ecology Series 5*, DOI 10. 1007/978-94-007-4244-7_4, Springer science+Busiiness Media B.V. 31-44

46. Pasche, N., Schmid, M.,Vazquez, F., Schubert, C., Wüest, A., Kesler J.D., Pack, M., A., Reeburgh, W.S., and Bürgman, H. (2011); *Methane sources and sinks in Lake Kivu.* J. Geophys. Res.116, G03006, DOI:10.1029/2011JG001690.

47. Patrice Stern and Jean-Marc Schottl (2019), *la boite à outil du consultant*, Ed. Dunod, collection BàO, 192p

48. Paul-Alain NANA and Moïse NOLA, (2020) ; *Lacs Monoun et Nyos au Cameroun : quand la nature, la science et les pensées autochtones africaines se confrontent*, LEF, pp76 - 81, 6p

49. R. Temdjim, P. Boivin, G. Ghazot, C. Robin, E. Roulleau (2004), *L'hétérogenéité du manteau supérieur à l'aplomb du volcan Nyos (Cameroun) révélée par les enclaves ultrabasiques,* C.R. Geoscience, 336, pp1239 - 1244

50. Reuchlin M., (1973 :3ème edition) ; *Les méthodes en psychologie*, P.U.F, Paris (France)

51. Sarmento, H., Isumbisho, M. and Descy, J.-P. (2006); *Phytoplankton ecology of Lake Kivu (Eastern Africa)* J. Plankton Res. 28(9): 815-829.
52. Sarmento, H., Isumbisho, M., Stenuite, S., Darchambeau, F., Leporcq, B. and Descy, J.-P. (2009); *Phytoplankton ecology of Lake Kivu (Eastern Africa): biomass, production and elemental ratios Verh.* Internat. Verein. Limnol. 2009, vol. 30, Part 5, p. 709-713
53. Tristan FERROIR (2009), *Eruptive dynamics*, January 14, 10p
54. W. Deuser, E. Degens, G. Harvey and M. Rubin (1973); *Methane in Lake Kivu: New Data Bearing on Its origin, Science,* vol. 181, no. 14094, pp51 - 54
55. Yves-Chantal GAGNON (2012 : 2ème edition), *L'étude des cas comme méthode de recherche,* Presse de l'université de Québec, 2012, 128p
56. Zhiwei Cui, Xinli Lu, Jialing Zhu, Wei Zhang (2020 :n°2) ; *Mesures dans le processus de dégazage du CO_2 en solution avec une référence particulière aux éruptions limniques provoquées par le CO_2,* proceedings, Géoscience, Tome 352, pp115- 126, 11p ;

2. PERIODICALS

57. Aurélien Augier (2021); *L'éruption de mai 2021 du Nyiragongo (République Démocratique du Congo) vue par interférométrie radar* ; Lycée Camille Guérin, Portiers, 10/06/2021, 6p
58. CEEAC and GFDRR (2021); *Les principaux dangers en Afrique Centrale*, Module, 183p
59. Dessus Benjamin, Laponche Bernard and Hervé Le Treut (2017) ; *Evaluer simplement l'importance pour le changement climatique des principaux gaz à effets de serre dans les scénarios mondiaux à partir des enseignements du dernier rapport du GIEC*, Article, 152, 15p
60. Gaju GAKINAHE, (1991) ; *L'eau et l'aménagement dans l'Afrique des grands Lacs,* : n°5, , Burundi (Bujumbura): Colloque de Bujumbura, Collection " Pays enclavés ", article.
61. Humanitarian Aid Office of the European Community (ECHO) (2003), *Le lac Kivu source de méthane, étude fondamentale, risques naturels. Exploitation du méthane du lac Kivu*, projet de station pilote, *data environnement*
62. Isumbisho Mwapu P., (2000); *Regime alimentaire des larves et juveniles de Limnothrissa miodon (Bonger, 1906) dans le lac Kivu, R. D. Congo,* mémoire D.E.S, Fac. Sc. Université de Liège, 40 pages
63. Michel Halbwachs, (2011), Exploitation optimale de la ressource en méthane du lac Kivu, Data environnement, 17 November 2011, 9pages
64. Muvundja Frabrice Amisi (2010); *Riverine nutrient inputs to lake Kivu;* Thesis, Makerere University, Reg.No. 2007/HD13/11068X, 92pages
65. Pietro Marzo, (2023); *Guide décolonisé et pluriversel de formation à la recherche en sciences sociales et humaines*, article
66. Stéphane Paquin, (2011) ; *Bouchard, Durkheim et la méthode comparative positive*, Québec (Canada), volume 30, n°1, article, 2011, 57 - 74, 19p
67. Tahitie Ben, (2010), *Le système de prévention et de gestion des catastrophes environnements au Cameroun et le droit international de l'environnement,* Master 2, University of Limoges, 64 pages

68. West Africa (1987); *The Monster of Lake Nyos*, Sydney, 1772 - 1773, 3p

3. UNPUBLISHED WORK

69. Auguy Blaise MUMBERE MAPENDO (2004); *Alerte sur la gestion des écosystèmes du Lac Kivu*, Goma (DRC), 42p

70. Dr BOUTELDJA FATHE, (2017), *Cours de géophysique appliquée*, Université 08 mai 1945, Guelma, 104p

71. Ephrem KAMATE KALEGHETSO (2018) ; *Pétrographie et géochimie des laves du volcan Nyiragongo (Nord-Kivu, R. D. Congo) : Influence de la viscosité sur les paramètres de propagation des coulées de laves menaçant la ville de Goma*, Master's thesis, Université de Liège, 77p.

72. Gildas Sylvère NGOMO (2016) ; *Comprendre le concept de la conscience en classe de philosophie au lycée: approche phénoménologique,* Master 2, Ecole Normale Supérieure, Libreville

73. Jacqueline FREYSSINET-DOMINJON, (1997) ; *Méthodes de recherches en Sciences sociales,* Paris (France), cours, AES, Montchrestien, 35p

74. Jean-Modeste MUSHIMIYIMANA (2020); *Middle Water Layer Temperature in Lake Kivu;* Kigali (Rwanda), African Institute of Mathematical sciences (AIMS), Award of Master of science in Mathematical sciences, 50p

75. Kaningini M, (1995) ; *Étude de la croissance, de la reproduction et de l'exploitation de Limnothrissa miodon au lac Kivu*, bassin de Bukavu (Zaïre). PhD thesis in science, P. UN, UNECES, Namur.

76. Klaus Tietze (1978) ; *Geophysikalische untersuchung des Kivusees und seiner ungewöhnlichen Methangaslagerstätte - Schichtung*, Dynamik und Gasgehatt des Seewassers, PhD, Thesis, Christian-Albrechts-Universität, Kiel, Germany, 150p ;

77. Monlay Rachid LAAMARI (2016-2017), *Cours de chimie organique semestre 2 SVI*, Université CADI AYYAD, 68p

78. S.A, (2013) ; *L'analyse critique,* Montmorency, cours, (340-102), 4p

79. Sekimonyo Shamavu Christian (2022) ; *Séminaire sur la répartition des blocs pétroliers et Gaz en République Démocratique du Congo*. University of Kinshasa: unpublished course, 30p

80. Sekimonyo Shamavu Christian, (2023) *Méthode de recherche scientifique*, course, ISDR/GL, Goma

4. DICTIONARIES AND ENCYCLOPAEDIAS

81. Universal Dictionary, 2010, 1551p

82. Larousse, Larousse junior dictionary, 2009

83. Larousse, Petit Larousse en couleurs, 1988

84. Le Robert, Le Grand Robert, Alain Rey, 2ème edition, electronic version, 2005

85. S.A, Dictionary of Geology

5. REPORTS AND OTHER DOCUMENTS

86. ARRICAU Victor (2020), *Géohistoire des risques naturels de trois lacs de cratère emblématiques du Massif Central français (lacs Pavin, Tazenat et Issarlès)*, Rapport de stage, Master 2, Université de Toulouse, 103p

87. Autorité du Bassin du Lac Kivu et de la Rivière Ruzizi (ABAKIR) (2020), *Etude de Base du Bassin du Lac Kivu et de la rivière Ruzizi,* Rapport, 250p

88. BIKUMU F. M., (2005); *La problématique du déficit énergétique dans la sous-*

région des Grands-Lacs africains, Pole Institute, 31p
89. Jean-Christophe SABROUX (2016), *La catastrophe du lac Nyos : de l'éruption limnique au dégazage contrôlé* ; RSN, CEA, Saclay, France, 18 January 2016, Amphithéâtre SOLEIL, 1p
90. Klaus TIETZE (2000), *Lake Kivu Gas Development and promotion-related issues: safe and environmentally sound exploitation,* Report, Republic of Rwanda, Ministry of Energy, Water and Natural Resources, 2000, Kigali, 110 p.
91. Kling George, Sally Maclntyre, J0rgen S. Steenfelt and Finn Hirslund; (2006); *Lake Kivu Gas Extraction,* Report on Lake Stability, Report n° 62721-0001, 103p
92. Ministère de l'Administration du territoire, (28 January 1994) ; *Rapport National des Activités liées à la décennie internationale de la prévention des catastrophes naturelles (DIPCN) au Cameroun*, Rapport, 9p
93. S.A. (2017), *the compass*, training module, 15p
94. TIETZE, K., and E. MAIER-REIMER (1977). *Recherches Mathématiques-Physiques pour la Mise en Exploitation du Gisement de Gaz Méthane dans le Lac Kivu (Zaïre/Rwanda).* Report No 76003, Bundesanstalt für Geowissenschaften und Rohstoffe, 2 Volumes, Hanover, 180 p.

7. WEBOGRAPHY

95. îhttp://svt.ac-creteil.fr/?Les-eaux-troubles-du-lac-kivu, consulted on 05/09/2023
96. îhttps://mhalb.pagesperso-orange.fr/kivu/fr, company Data environnement, Michel Halbwachs
97. îhttps://static.fnac-static.com/multimedia/editorial/pdf/9782381820064.pdf, consulted on 06/09/2023
98. îhttps://www.scribbr.fr/methodologie/article-scientifique/discussion-article-scientifique/, published by Justine Debret, accessed 22/09/2023
99. îhttps://www.scribbr.fr/methodologie/collecte-de-donnees/, published by Gaspard Claude, accessed 17/09/2023
100. ↑www.chem.qmul.ac.uk, Queen Mary University of London, consulted on 21/06/2023
101. ↑www.hydrocarbures.gouv.cd, DRC Ministry of Hydrocarbons, consulted on 16/08/2023
102. Îwww.reflexions.uliege.be/cms/c 340052/en.les-secrets-du-lac-kivu, review published by Philippe Lecrenier, Université de Liège, consulted on 20/06/2023
103. ↑www.hanhimanti.com, published by Hanhimanti, accessed 17/09/2023
104. îhttps://geo-ocean.fr, published on 06 May 2022 and accessed on 23/09/2023
105. ↑www.larousse.fr, consulted on 16/06/2023 at 2.20pm
106. ↑www.lerobert.com, consulted on 16/06/2023 at 2:35 pm
107. îhttps://inis.iaea.org, consulted on 20/08/2023
108. twww.researchgate.com, consulted on 20/08/2023
109. twww.fnac.com, Alexandra Midael, 2009, consulted on 25/08/2023 at 23:50
110. ↑www.natura-sciences.com, consulted on 06/09/2023
111. ↑www.revues.cienceafrique.org, consulted on 07/09/2023
112. thttps://kivu-power.com, Kivu Power, consulted on 20/09/2023
113. thttps://ground.geophysicspr.com, consulted on 23/09/2023 at 8:40 am.
114. thttps://ressources.arcgis.com, consulted on 23/09/2023

115. thttps://web.maths.unsw.edu.au, University of New South Wales, accessed on 23/08/2023 at 11:23 a.m.
116. thttps://gps.gov/systems/gps/, consulted on 23/09/2023 at 11:27.
117. thttps://businessyield.com, accessed on 23/09/2023 at 12:06pm.

APPENDIX

Appendix 1. Figure 14. African lakes susceptible to limnic eruption

Source: Dangers inconcevables, p17. <u>(https://static.fnac-static.com/multimedia/editorial/pdf/9782381820064.pdf,</u> consulted on 06/09/2023) : Meromictic lakes in orange

Appendix 2. Figure 15: Areas at risk of limnic eruption

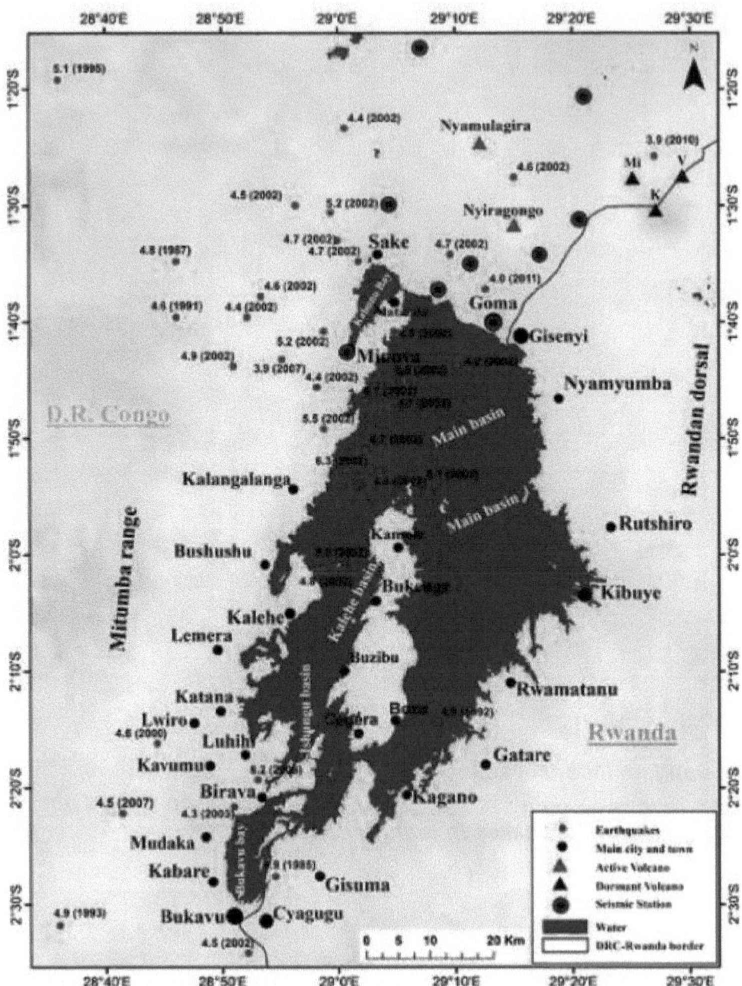

Source: ***<u>www.researchgate.com,</u>*** ***accessed on 20 August 2023 at 21:00.***

SURVEY QUESTIONNAIRE

As part of our dissertation, we are carrying out a study on ***"The problem of gas from Lake Kivu: a comparative study with the limnic eruption of Lake Nyos in Cameroon in 1986"***. It is with this in mind that we are asking you to kindly answer these questions in the interests of science.

We are therefore asking you to help us by answering the questions below faithfully and honestly.

We guarantee your anonymity.

Instructions :

- Put a cross in the box corresponding to your answer.
- Fill in the dotted lines with your own ideas.

I. IDENTITY

1. Your gender M F 2. Your age a) Under 18 b) 18 - 25 c) 26 - 30 d) 31 - 35 e) 36 - 40 f) 40 and over 3. Your marital status: a) Single b) Married c) Widowed d) Divorced 4. Your level of education: a) Primary b) Secondary c) Graduate d) Degree

II. QUESTIONS PROPER

1. Have you heard of the limnic eruption?

a) Yes I I b) No I

2. Is an limnique eruption possible in Lake Kivu?

a) Yes I , I b) No I I

If yes, give reasons and if not, give reasons

3. What would be the consequences (positive and negative) of the limnic eruption of Lake Kivu?

a) Death of men and livestock from asphyxiation
b) Greenhouse gas emissions
c) Population displacement
d) Ecological imbalance
e) Respiratory disease and burns
f) Loss of wealth for the country

4. How can the risks and disasters of the limnic eruption of Lake Kivu be managed and prevented?

4.1. How do you manage risk?

a) Move the population away from the high-risk area
b) Study the probability of a limnic eruption ^Z^Z
c) living in high-altitude areas ^Z^Z
d) prohibit agricultural and pastoral activities in high-risk areas

4.2 How can disasters be managed? a) Isolate the area affected by the gas fumes b) Set up a disaster relief management structure in the surrounding towns c) Evacuate the injured d) Try to find the survivors e) Set up a crisis unit f) Set up a disaster assistance committee g) Evacuate the population from the danger zone h) Maintain "toxic danger" signs i) Find a remote resettlement area j) Other to be specified .... 4.3. How can risks be prevented? a) Drain gas from Lake Kivu b) Install seismic stations in the lake c) Inform the surrounding population of possible risks

4.4 How can the disaster be prevented? a) Set up the Lake Kivu degassing project b) Extract the methane gas c) Regularly monitor the lake's limnological state

5. What are the comparative endogenous, exogenous and heterogeneous factors between Lake Kivu and Lake Nyos with regard to the limnic eruption?

5.1. Endogenous factors a) Convergent a. Both lakes are meromictic b. High concentrations of dissolved gases c. Fed by volcanic gas d. High depth b) Divergent e. The gas contained in Lake Kivu is a thousand times greater than that in Nyos before the disaster f. Lake Kivu is fed by volcanic fractures whereas Nyos is in a crater g. Other ....

5.2. Exogenous factors a) Convergent a. Proximity to volcanic areas b. Earthquakes b) Divergent c. Landslides d. Surrounding population e. Volcanic lava flows 5.3. Heterogeneous factors a) Convergent a. Imbalance in Planck tectonics b. Volcanic activity c. Earthquakes b) Divergent d. Internal volcanic eruption e. Climate change

Printed by Books on Demand GmbH, Norderstedt / Germany